D0906087

MANNED SPACECRAFT

The Pocket Encyclopedia of Spaceflight in Color

Other books by Kenneth Gatland:
MISSILES AND ROCKETS
ROBOT EXPLORERS

By Philip Bono and Kenneth Gatland:
FRONTIERS OF SPACE

MANNED SPACECRAFT

by
KENNETH GATLAND
FRAS, FBIS

Illustrated by
JOHN W. WOOD
TONY MITCHELL
JACK PELLING

Second Revision

MACMILLAN PUBLISHING CO., INC.
New York

Link House, West Street, Poole
Dorset BH15 1LL

Macmillan Publishing Co., Inc.
866 Third Avenue, New York, N.Y. 10022

Library of Congress Cataloging in Publication Data

Gatland, Kenneth William, 1924–
Manned Spacecraft

(The Pocket Encyclopedia of Spaceflight in Color)
(Macmillan color series)
Includes index
1. Astronautics - Popular works. 2. Manned Space Flight. I. Title
TL793.G34 1976 629.47 76-1994
ISBN 0-02-542820-9

First American Edition 1967
First American Edition of Second Revision 1976

Printed and Bound in Great Britain

ACKNOWLEDGEMENTS

As a vice-president of the British Interplanetary Society it has been my pleasant duty to meet leading personalities concerned with space technology of both the United States and the Soviet Union, and to discuss with astronauts and cosmonauts their unique experiences of orbital flight.

The results of some of these encounters are given in the following pages where the attempt has been to interweave the human experience of spaceflight with the considerable technical achievement which the conquest of space represents. Opportunity has also been taken to sketch the steps that lie immediately ahead in manned spaceflight, including the development of orbiting laboratories.

A special feature of the present volume is the colour illustrations, some of which, like Mercury, Vostok, Apollo and Soyuz, have been prepared after examination of actual hardware, and others from detailed drawings and photographs supplied by companies concerned with project development. The principal artist, John Wood, has worked with me for the past 25 years (following our association in the design department of Hawker Aircraft Limited at Kingston-upon-Thames) and is now among the best known in the aerospace field. He was assisted in the present work by Tony Mitchell, another Hawker protégé, whose own talents for aerospace illustration are fast becoming recognized.

Special mention must also be made of Rockwell International artists whose excellent series of impressions of the Apollo lunar mission appear on pages 77–81, 83, 84, 87 and 91. My good friend W. A. 'Bill' Dunn of the United States Information Service in London supplied the transparencies showing actual US spacecraft, the Kennedy Space Center and astronaut photographs taken from orbit. The publishers join me in expressing special appreciation to both these sources.

Particular thanks are also due to the following for photographs and technical information: McDonnell Douglas (Mercury, Gemini and ATDA); Rockwell International (Space Shuttle,

Apollo command, service and adapter modules); Grumman Aircraft Engineering Corporation (Apollo lunar module); The Martin Company (Titan 2); Lockheed Aircraft Corporation (Agena target vehicle); General Electric Company of the USA (ALSEP); and Teledgne Ryan Company (Apollo landing radar). The comparative drawings of Salyut space-stations (page 236) were based on work by Sven Grahn.

Once again it is a pleasure to record appreciation of the flow of technical material from the National Aeronautics and Space Administration in Washington, and the ready assistance offered by NASA to researchers in this field. Definitive information on Soviet space progress is characteristically more difficult to obtain; but even here there has been evidence of improved lines of communication. I am particularly indebted to Mrs Iris Smith of the Novosti Press Agency in London for assistance in obtaining photographs and data relating to Soviet space technology.

All launch and orbital data are those officially released by the country of origin. Where orbital parameters are given they normally refer to the initial orbit. My earlier books, *Spacecraft and Boosters*, Vols. 1 and 2 (Iliffe Books Ltd), detailing space launchings of 1961 and 1962, have been helpful in arriving at qualified data. The following additional sources have been freely consulted: *Flight International, Aviation Week & Space Technology, Spaceflight* (a monthly publication of the British Interplanetary Society), and *Space Log* (edited by Thomas L. Branigan for TRW Space Technology Laboratories).

Lastly, much is owed to my wife Doreen whose unstinting efforts in typing the original manuscript contributed greatly to the ability to produce the text in little more than two months.

For the convenience of international editions the 80-page colour section is bound at the front of the book. The captions which appear on the colour pages are amplified in Chapter Eight where full details of the subjects are given with appropriate reference numbers.

KENNETH W. GATLAND

CONTENTS

INTRODUCTION

Money spent on the advancement of knowledge is seldom wasted. Given time, the investment nearly always reaps a rich reward, often in the most unexpected directions.

Astronautics – the science and technology of spaceflight – is likely to be no exception. Combining as it does almost every scientific and technical discipline, it has already become a 'spear-head technology' stimulating the development of new materials and processes of manufacture. All the time, because of space activity, there is a steady advancement in such basic fields as metals, ceramics, plastics, micro-electronics, power generation, industrial management, and many more. The effects on industry in general are evolutionary and largely unseen, but for all that they may be expected to have the widest significance for twentieth-century man.

There may even be important medical benefits. The close attention now being paid to the work of the human body under space conditions is resulting in more data being amassed on vital physiological functions – particularly concerning the circulatory system under stress – than in decades of conventional medicine. Techniques learnt in space, such as the monitoring of bodily functions by telemetric sensors (e.g. heart-rate, pulse, and respiration), are starting to revolutionize medical instrumentation in hospitals. Semi-automated hospitals, where a patient's physiological reactions are automatically reported to master consoles, are already in prospect which both alert the medical staff and provide computerized data on the patient's condition.

However, new and perhaps even more exciting opportunities are about to open, for medical research could benefit from the close attention that space scientists are paying to the living cell in terms of the search for life on other planets. The automated biological laboratory, which the American Viking spacecraft is to land on Mars in 1976, will have both soil sampling and air sampling apparatus for checking on the presence of micro-organisms and reporting findings to Earth.

In this area of advanced biology the interests of space science and medical research may merge to the benefit of mankind, for a greater understanding of the properties of the living cell is fundamental to winning the battle against malignant disease. It is interesting and vitally significant that important experiments concerning the structure of living matter are now taking place in research centres devoted to space problems.

If these are the 'hidden' benefits of space research, much more is nakedly apparent. Indeed, the most obvious is already taken for granted in terms of satellites that now allow television to span continents and oceans. The development of satellite communications in the first decade of the space age has been little short of miraculous; and when we see programmes on our TV screens relayed across the Atlantic we little think of the small box of high-grade electronics working unattended in the hostile environment of space that makes it all possible. Even Arthur C. Clarke, a former chairman of the British Interplanetary Society, who first proposed the synchronous-orbit TV satellite in 1945, could scarcely have imagined the standards of reliability now being achieved in space electronics.

The achievement, of course, extends far beyond communications. Satellites warning of the approach of hurricanes and typhoons – as well as performing routine global weather survey – are in regular operation. Others serving as 'radio-stars' for the precise navigation of ships and aircraft have been demonstrated, and the future is bright for a combination of these systems in the form of multipurpose spacecraft.

If all this has been possible in the first ten years of space technology, what will twenty or thirty years of development have in store? It is not difficult to predict that our planet will have a network of space communications serving the interests of governments, trade and industry. Computer-controlled information banks, flashing the most up-to-date data of all kinds by satellite to subscribers around the world, have already been suggested, which may offer immense opportunities for commerce and education.

One other important development can also be forecast. This

is the direct-broadcasting satellite poised in geo-stationary orbit beaming radio and television directly to domestic aerials. Such powerful satellites, dispensing with costly ground terminals and cross-country relay stations, will depend on cheaply produced dish aerials pointed skywards to intercept the signals coming to them directly from space. In this way programmes beamed to a satellite from a single ground transmitter will be re-transmitted over a vast area.

Apart from the utility of such satellites in developed parts of the world, they could become one of the most effective methods of bringing education to community centres in the emergent countries.

The growth of supersonic air travel, first over the Atlantic and then the Pacific, is also likely to depend heavily on satellites for air-traffic control, ensuring the maximum of safety for the international commuter.

None of these techniques demands radical advances from systems already developed in the first space decade, and their influence is likely to be much wider than our present imaginings.

But even this is not the limit of the space age potential. Those who strove to build the science of astronautics in many countries, who were filled with the desire to promote a deeper understanding of the Universe through interplanetary exploration, have always regarded themselves as members of a greater international family. It is therefore heartening that space activity, rather than adding to the human burden in terms of armaments, is actually helping to reduce tensions.

In a very real sense space travel is sweeping aside national boundaries, making closed societies more difficult and breaking down national distrust. Eventually, the space race that grew out of the East–West rivalries of the 1950s must give way to collaboration; and it may well promote the climate for at least a degree of disarmament.

Already the United Nations treaty on outer space, signed in 1967 by countries of both East and West, bans weapons of mass destruction from outer space and provides for equal access to the Moon and other celestial bodies by all nations. Out of this could grow collaboration in the exploration of the Moon. When it becomes technically possible, in principle there is nothing to prevent the United States and the Soviet Union, under an agreed programme, combining in a joint logistics operation to carry prefabricated parts to the Moon for the development of an international lunar base.

A lead in this direction, in fact, was given by the late President Kennedy. Addressing the U.N. General Assembly on 20 September 1963, he said: 'Why . . . should man's first flight to the Moon be a matter of national competition? Why should the United States and the Soviet Union, in preparing such expeditions, become involved in immense duplications of research, construction and expenditure? Surely we should explore whether the scientists and astronauts of our two countries – indeed of all the world – cannot work together in the conquest of space, sending some day in this decade to the Moon not the representatives of a single nation but the representatives of all our countries.'

Although the suggestion lacked a formula for combining East and West space development resources, it expressed the feelings of many people. In an atmosphere of reducing international tensions, at some future stage, it may be possible to achieve a degree of technological co-operation; and possibly the best chances lie in the direction of expensive projects involving logistics supply – the lunar base and manned interplanetary exploration.

As a first step – as Colonel John Stapp has proposed to the International Academy of Astronautics – it might be possible to develop an international orbiting laboratory. This would not involve the assembly of a single space-station in Earth-orbit but rather the orbital grouping of a number of spacecraft in conjunction with a 'dormitory' module. The various components, put up from launch centres in both the United States and the Soviet Union, would serve different functions. For example, one could work in astronomy, another in biology and another in meteorology and Earth-observation. And each would benefit from the others in terms of logistics support, equipment repair and astronaut safety. Astronaut-scientists would move between the different vehicles in a kind of 'space-taxi' using techniques of docking already developed in the Gemini and Apollo programmes. As will be seen in Chapter Four, the modular assembly of an orbiting laboratory is being investigated in the Apollo Applications Programme.

The prospects of this type of development are considerable. While scientific satellites and space probes have begun to revolutionize our understanding of the Universe, cameras and spectral sensors turned on the Earth itself have shown the promise of new opportunities for assessing the world's natural resources. The National Aeronautics and Space Administration has defined more than 200 Earth-related experiments that could be performed from a manned orbiting laboratory including cartography (mapping), geology, oceanography, water management, ice survey, agriculture, meteorology and other studies unique to the space environment.

Dr Wernher von Braun, technical director of the Marshall Spaceflight Center in Huntsville, Alabama, has written enthusiastically of these prospects. Everywhere he goes, man cuts trees, tills soil, builds houses, factories and roads. All this random activity is detectable from space. This information collated with Earth-based information can be used to assess the world's burgeoning population and future needs. From the high-ground of space it may be possible to detect crop disease. Black stain rust, von Braun points out, is difficult to detect in its early stages. Remote sensors can spot the rust several days earlier than a man who is standing on the ground.

Using similar remote techniques it should be possible to discover water and mineral imbalance in the soil, leading to better use of land in agriculture. By measuring minute differences in soil temperature it may be possible to detect underwater rivers, or measure snowfall and spring thaws helpful to the management of water in storage lakes. Life patterns in the seas, and feeding grounds for fish, may also be discernible.

Many of these remote sensing techniques are possible because of the prior development of military reconnaissance and surveillance satellites. Observations made by the astronauts themselves have contributed important information.

When I discussed the concept of the international orbiting laboratory with Dr von Braun, he thought it inconceivable that such unique opportunities for research should be restricted to scientists of the richer nations. He envisaged the possibility of grants being established on the pattern of Antarctic exploration, whereby graduate scientists of all countries could qualify, after special training, to conduct research aboard space vehicles.

PREFACE TO THIRD EDITION

When the above was written in 1967 the potential of Space was still largely unknown. Now 12 men have left footprints in the ancient dust of the Moon and three teams of Skylab astronauts have made their home in orbit aboard an 85-ton space station – for 170 days!

Progress has been fantastic. Robot spacecraft have been sent out to the planets Venus, Mercury, Mars, Jupiter and Saturn. And the search for life on Mars has begun.

But space research is more than exploration. Satellites for all kinds of practical purposes have emerged as we knew they would – spinning a web of communications around our planet, beaming education to isolated communities by television, helping to fight illiteracy, disease and over-population. The Earth resources satellites as well as reporting on the health of agricultural crops and forests and spotting pollution are helping us find minerals and fossil fuels.

In the meantime, the value of man-in-space continues to grow. Research already carried out in Skylab suggests that we shall see space factories using the weightless environment to make new high-strength/low-weight materials, grow super-crystals for the electronics industry and purify vaccines. In the 1980s scientists will commute to and from orbit in the revolutionary Space Shuttle – part rocket, part aircraft – which takes off vertically and returns to land on a runway.

No less impressive has been the progress towards space co-operation achieved in the US/Soviet experiment to dock together in Earth-orbit Apollo and Soyuz spacecraft and the joint training of their crews in both countries.

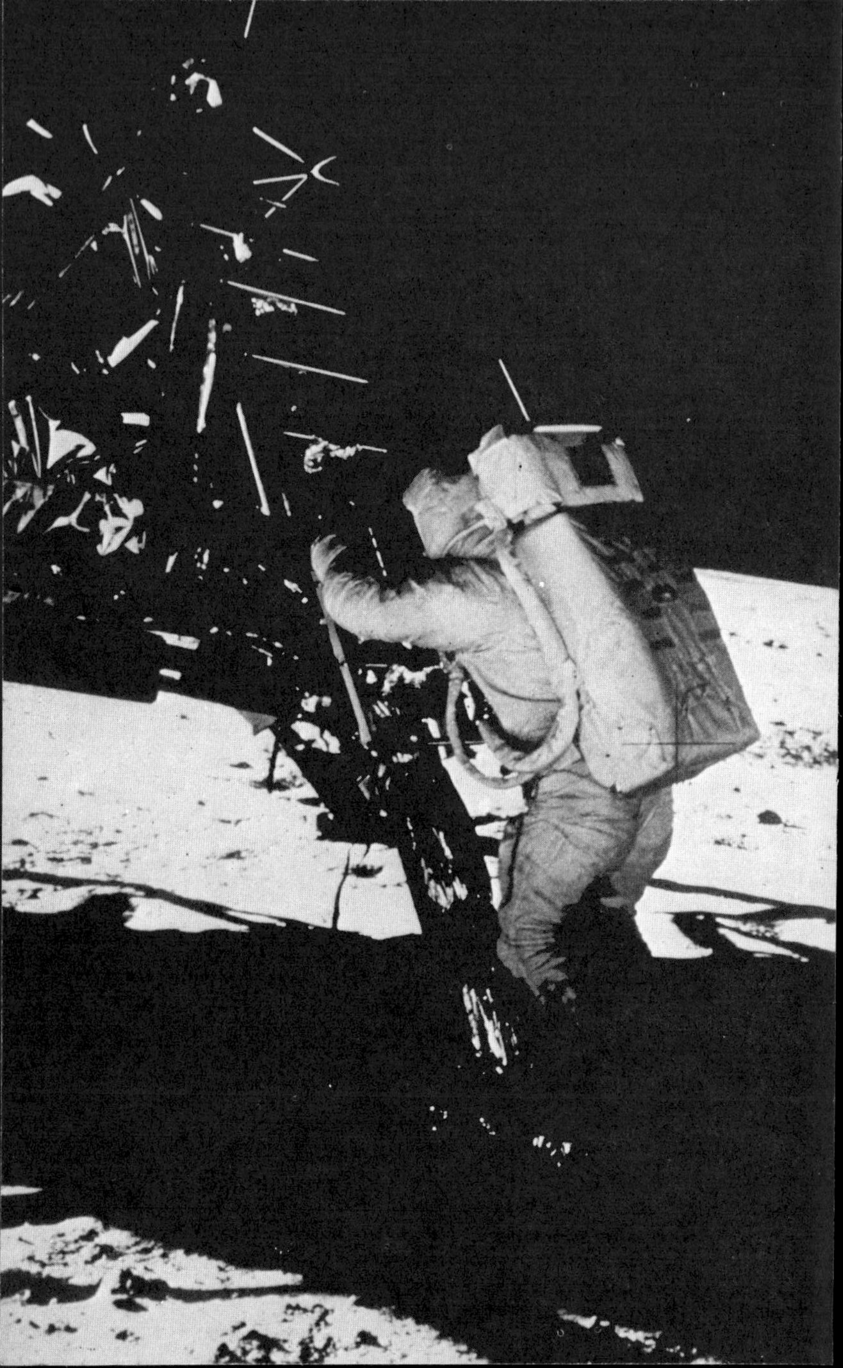

FIRST MEN ON THE MOON. *(Left)* Apollo 11 Lunar Module Pilot Edwin Aldrin is photographed by Neil Armstrong as he descends to the Moon's surface, Tranquillity Base, 21 July 1969. *(Right, top)* Aldrin poses for an historic picture. Mirrored in his helmet visor are photographer Neil Armstrong, the Lunar Module and part of the scientific package. *(Inset)* Metal plaque on the forward leg of the Lunar Module's descent stage (which remained behind on the Moon) reads:

'Here, men from the planet Earth first set foot upon the Moon, July 1969 A.D. We came in peace for all mankind.'

The metal plaque also bears the names of the three astronauts, Neil Armstrong, Edwin Aldrin and Michael Collins (who monitored the landing operation from lunar orbit in the Apollo parent). The name of President Nixon is also inscribed.

The Colour Pages

The 80 colour pages which follow illustrate United States and Soviet manned spacecraft, launch vehicles, launch sites, spacesuits, views from orbit, and the Apollo lunar mission. The reference number for each subject corresponds to the appropriate text matter. An index appears on pages 300–4.

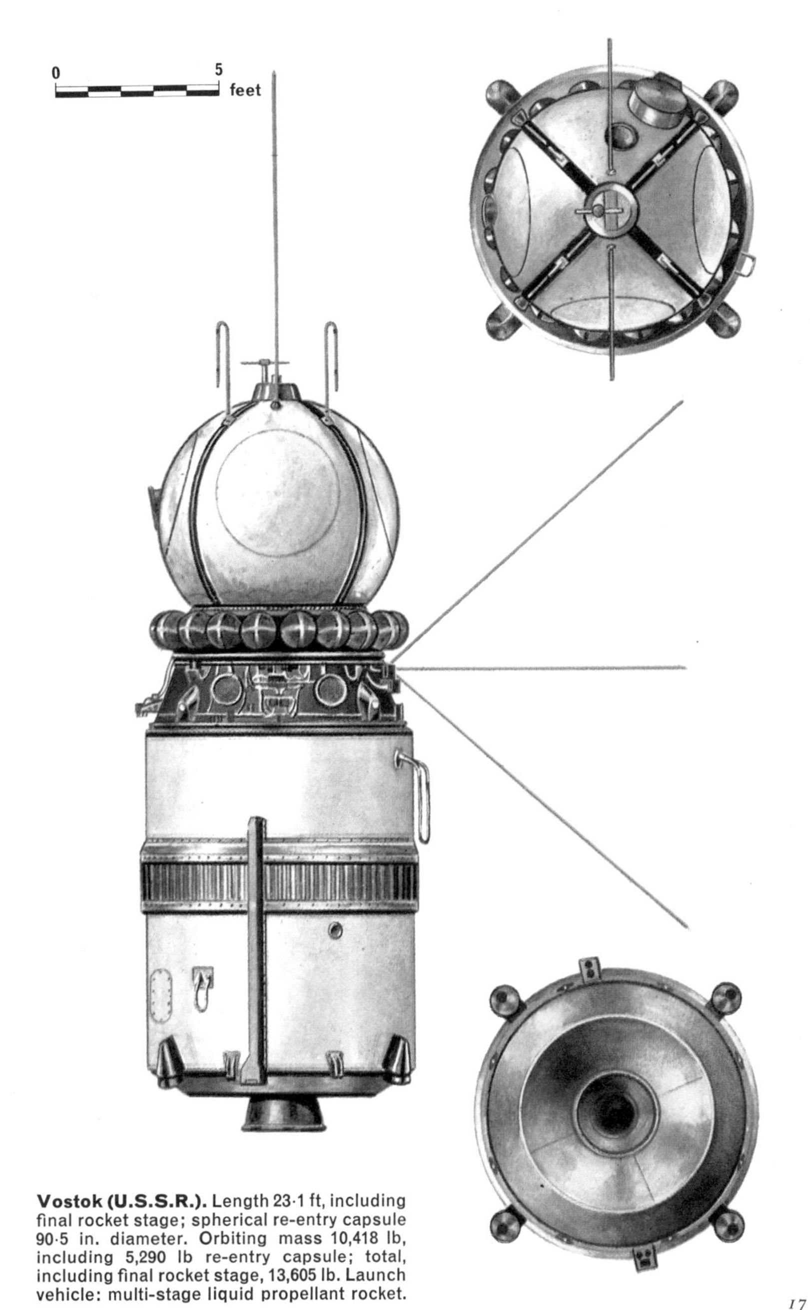

Vostok (U.S.S.R.). Length 23·1 ft, including final rocket stage; spherical re-entry capsule 90·5 in. diameter. Orbiting mass 10,418 lb, including 5,290 lb re-entry capsule; total, including final rocket stage, 13,605 lb. Launch vehicle: multi-stage liquid propellant rocket.

First three Soviet satellites to scale. Silhouettes indicate vehicles to same scale as spacecraft on facing page.

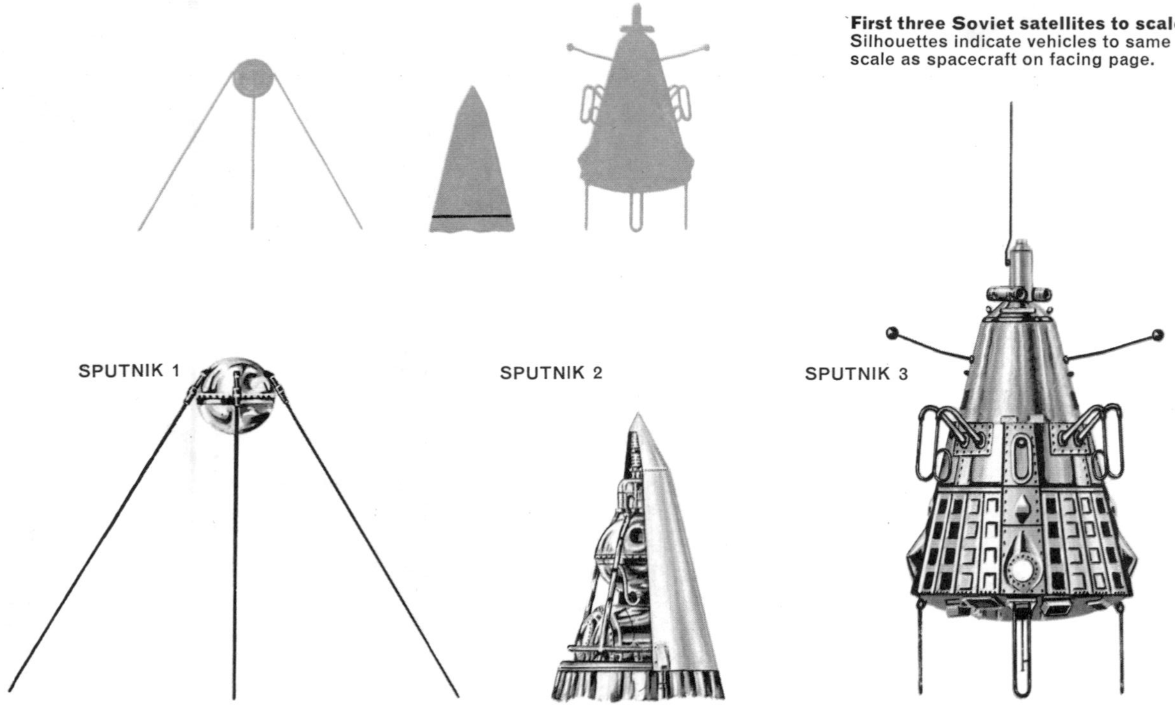

SPUTNIK 1

SPUTNIK 2

SPUTNIK 3

Vostok and Voskhod spacecraft complete with jettisonable nose fairings and final stage rockets. Note similarity between final stages of Lunik and Vostok launchers. (Voskhod drawing is provisional.)

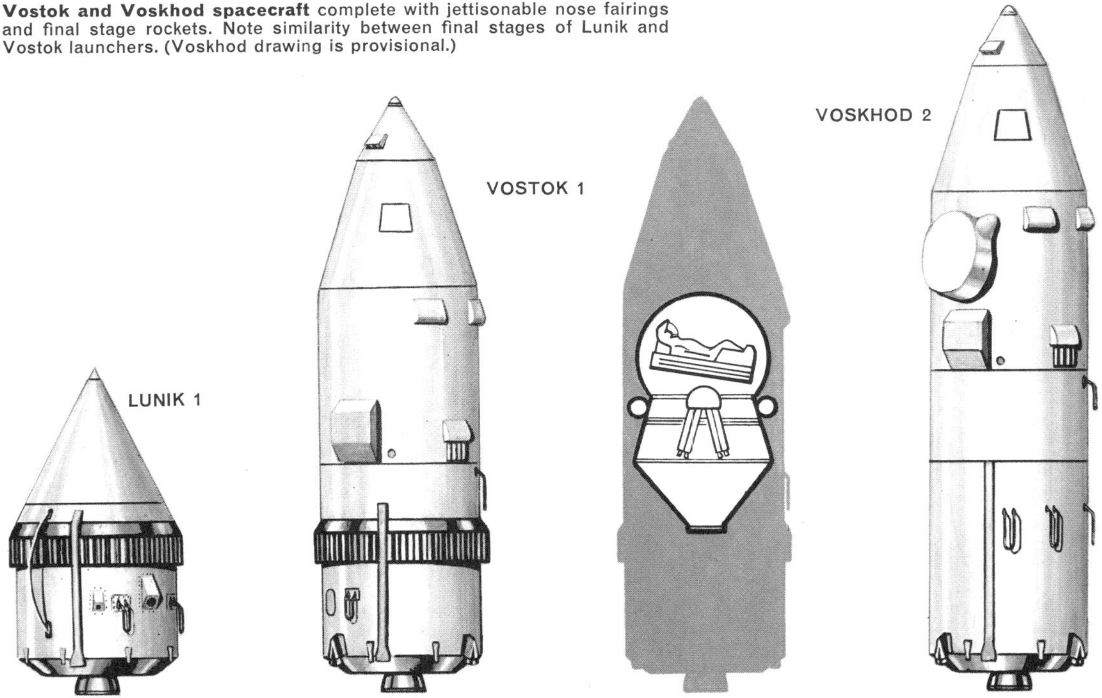

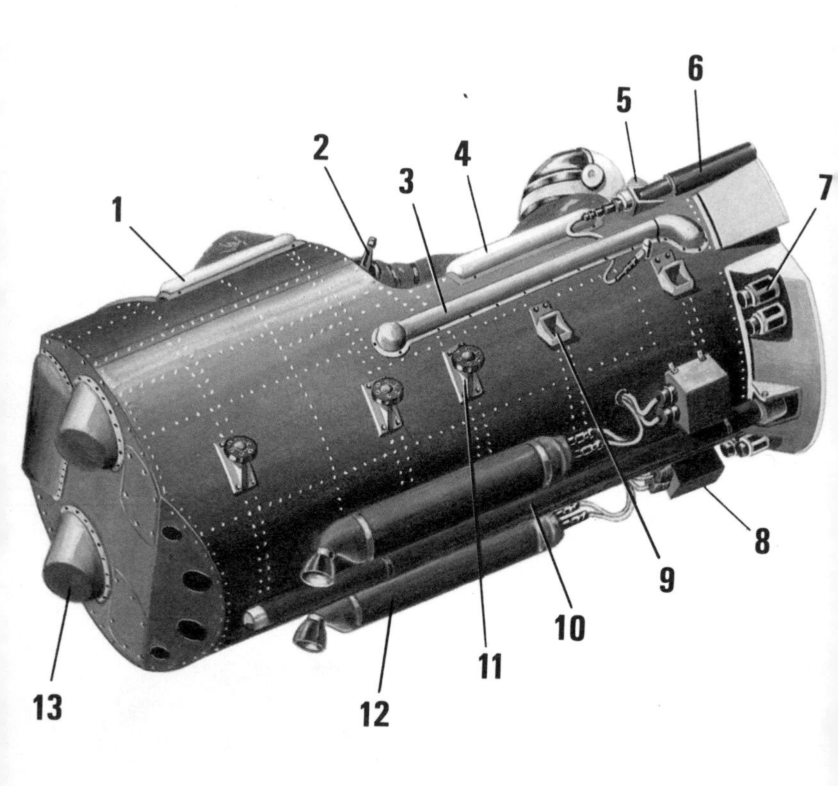

(Above) **Vostok ejection seat.** *Key:* 1. Leg pad; 2. Handle for shoulder harness; 3. Conduit; 4. Arm pad; 5. Socket connector; 6. Back plate ejector; 7. Aneroid pressure sensors; 8. Radio beacons; 9. Seat guides; 10. Piston; 11. Seat runners; 12. Ejection rockets; 13. Seat buffers.

(Right) **Vostok escape hatch.** *Key:* 1. Nose fairing; 2. Fairing aperture; 3. Capsule support flange; 4. Hatch flange; 5. Cosmonaut's ejection seat; 6. Re-entry capsule; 7. Equipment module containing retro-rocket; 8. Final rocket stage.

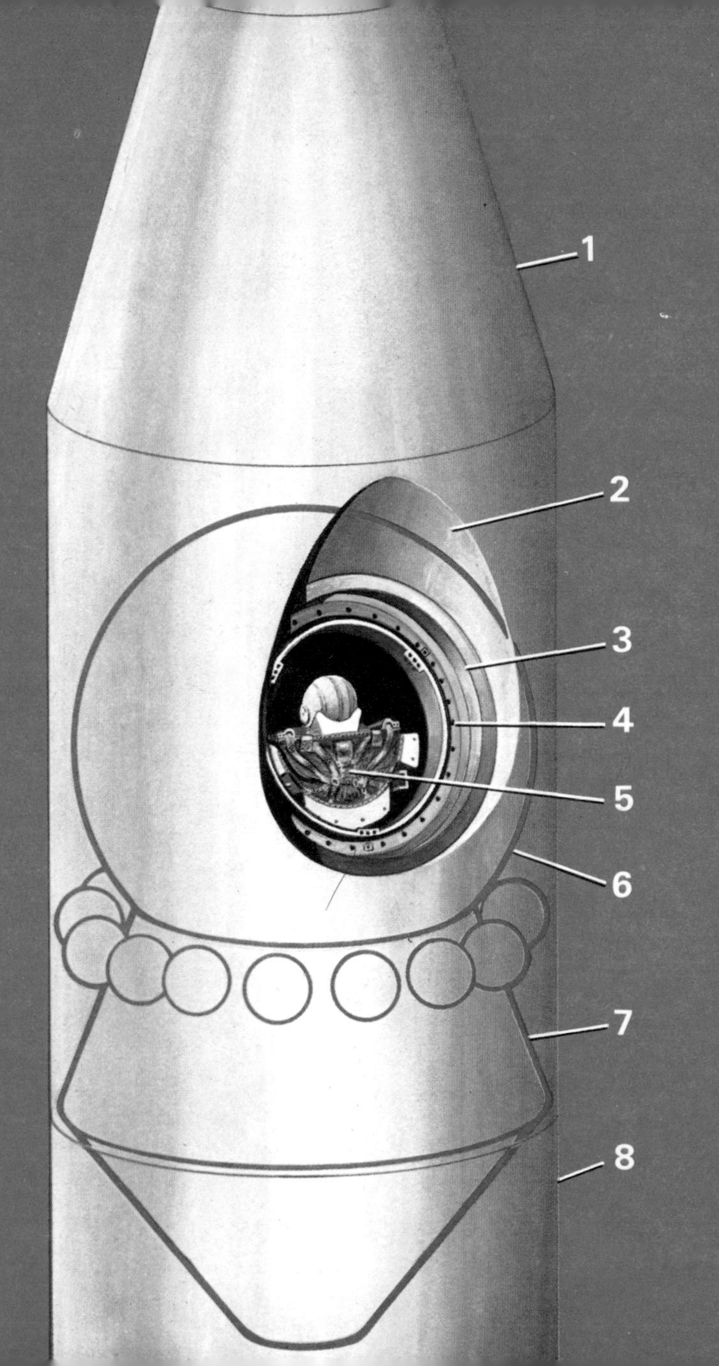
1
2
3
4
5
6
7
8

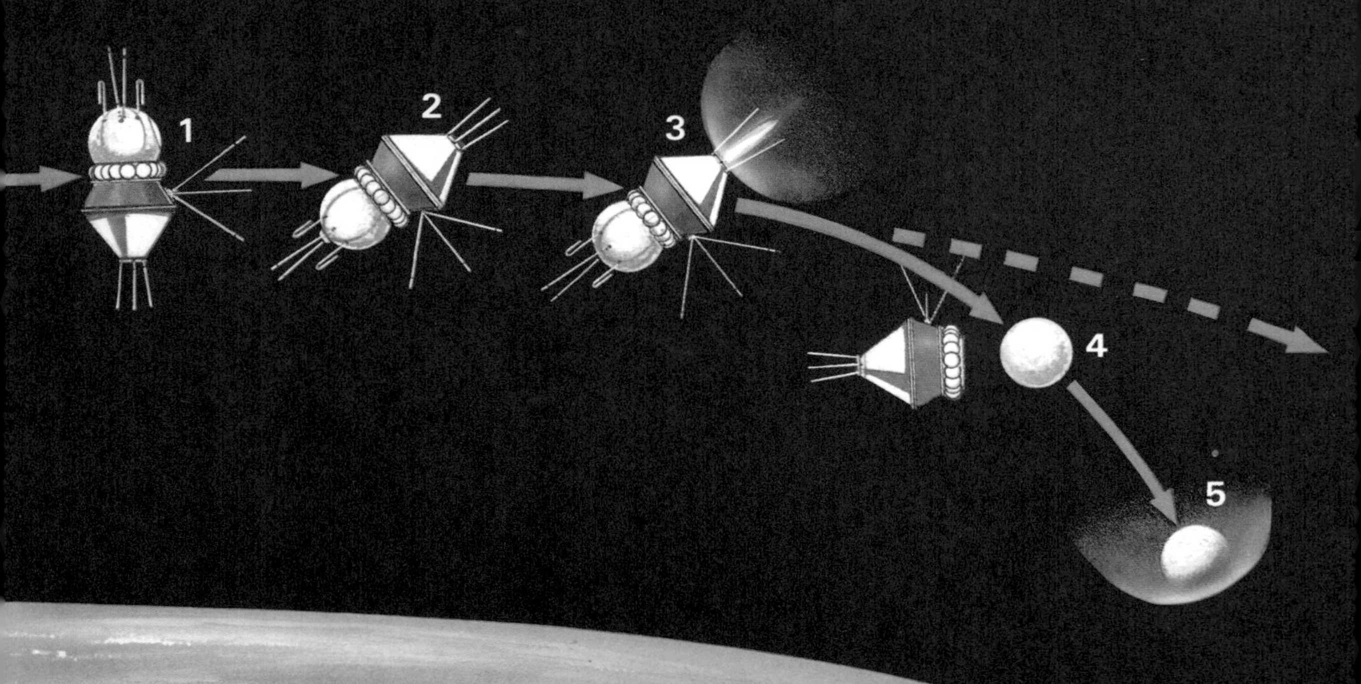

Vostok re-entry. *Key:* 1. Spacecraft in orbit; 2. Orientated with solar cells towards sun; 3. Retro-fire; 4. Capsule separates; 5. Re-entry heating.

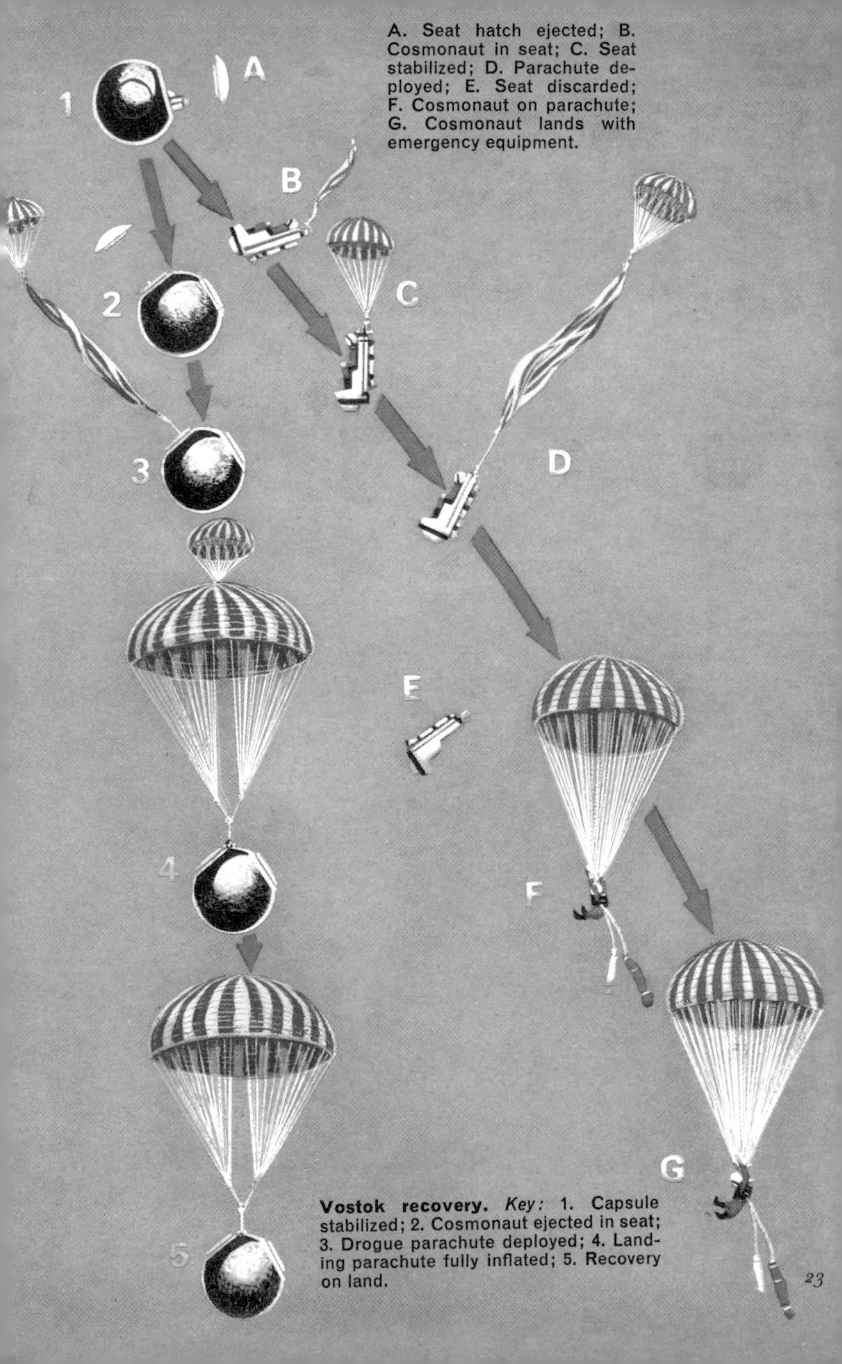

Vostok recovery. *Key:* 1. Capsule stabilized; 2. Cosmonaut ejected in seat; 3. Drogue parachute deployed; 4. Landing parachute fully inflated; 5. Recovery on land.

19
20
21
22
23
24
1
2
3
4

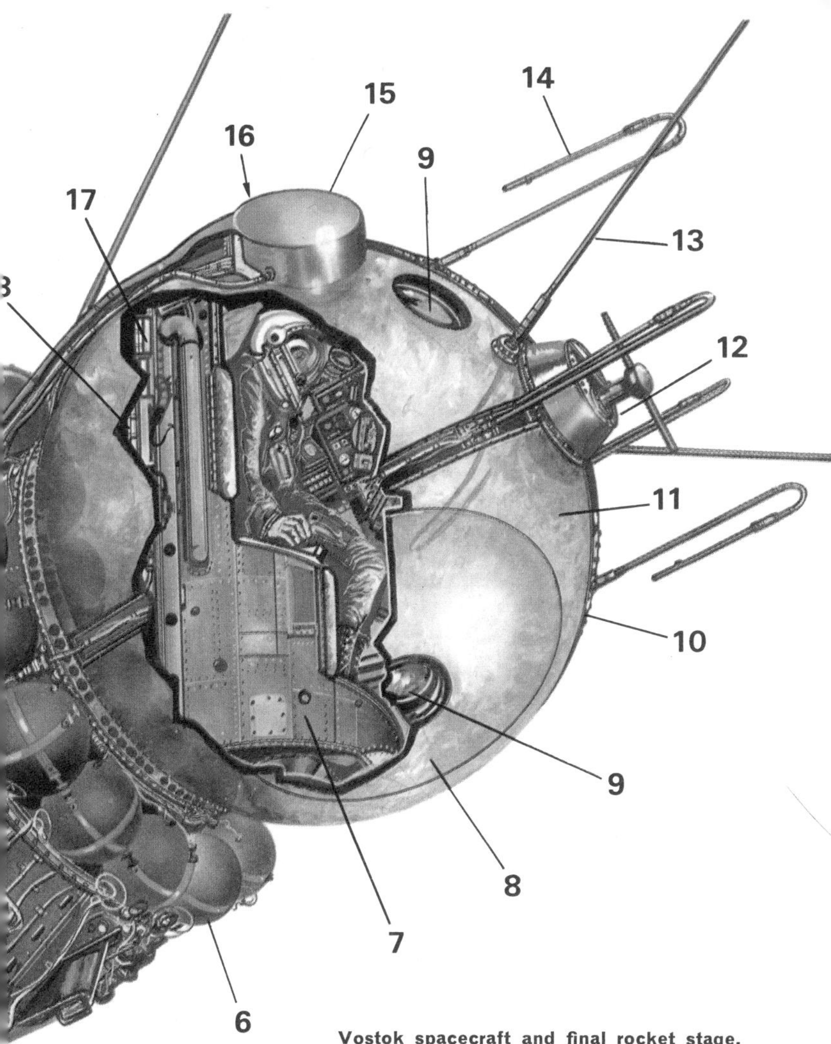

Vostok spacecraft and final rocket stage. *Key:* 1. Attitude control motors (4); 2. Interstage attachments (8); 3. Final stage rocket; 4. Access hatch; 5. Vernier nozzles; 6. Oxygen and nitrogen storage bottles (16); 7. Cosmonaut's ejection seat; 8. Equipment inspection hatch; 9. Portholes; 10. Tensioning bands holding re-entry capsule; 11. Spherical re-entry capsule with ablative heat shield; 12. Electronics package; 13. Whip aerials 11 ft long (2); 14. Control command aerials (4); 15. Multiplex connector; 16. Ejection seat hatch (unseen at rear); 17. Ejection seat rails; 18. Ejection seat rocket motors; 19. Whip aerials (4); 20. Electrical harness; 21. 'Paper-clip' command aerial; 22. Rocket stage extension (shrouding retro-rocket of equipment module); 23. External conduit; 24. VHF aerial.

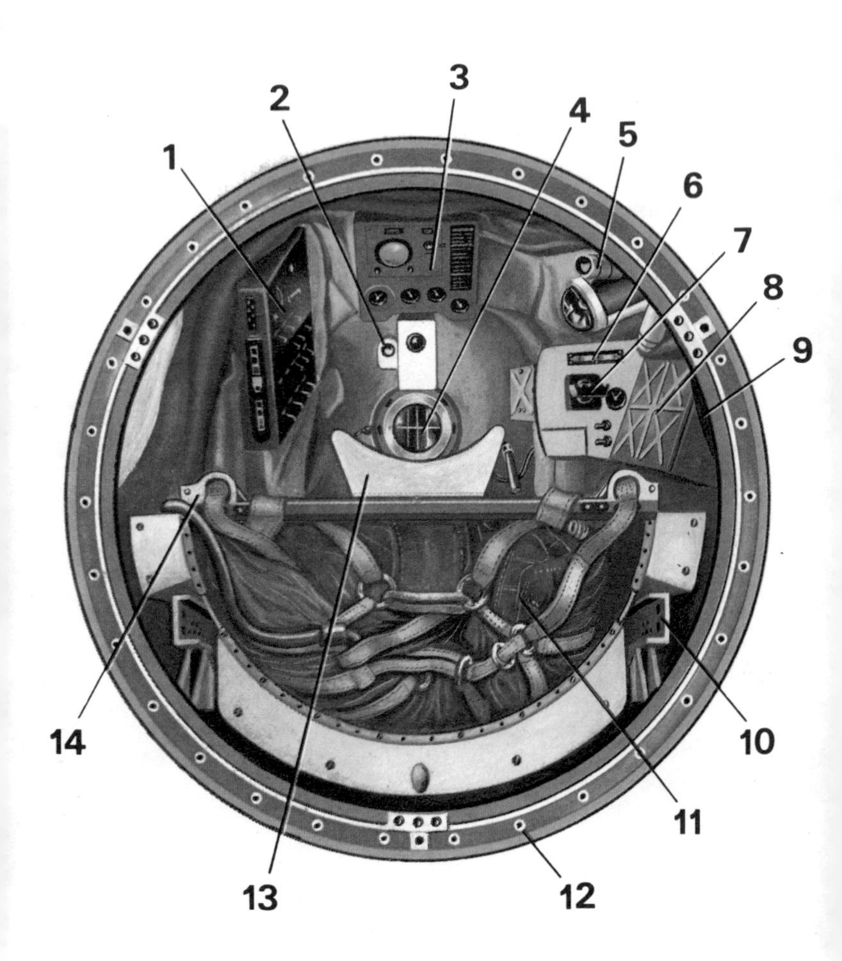

Vostok interior (as seen through ejection seat hatch). *Key:* 1. Switch panel; 2. Television camera; 3. Instrument panel with revolving Earth-globe; 4. Porthole, incorporating 'Vzor' optical orientation device; 5. Mirror; 6. Radio; 7. Manual control handle; 8. Food container; 9. Multiplex connector; 10. Ejection seat rail; 11. Seat drogue and parachute; 12. Threaded sockets for hatch cover; 13. Ejection seat headrest; 14. Anchorage for parachute harness.

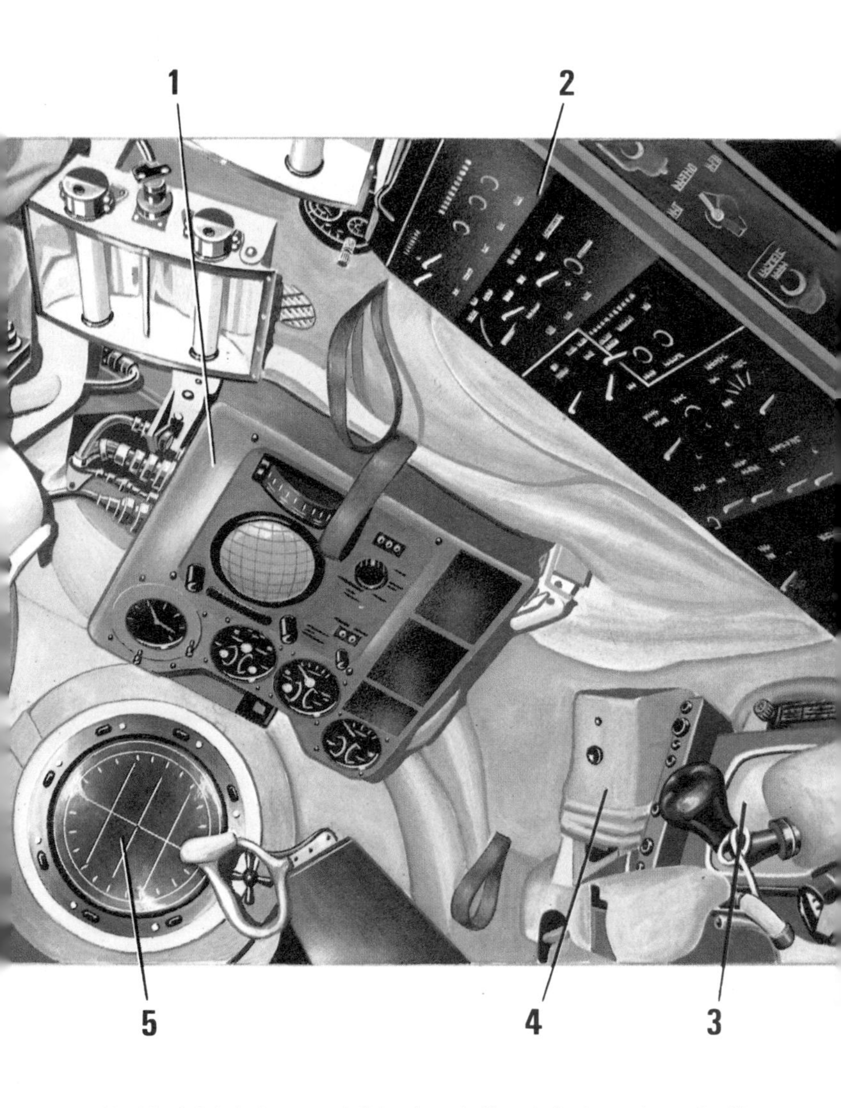

Voskhod 2 interior, near left-hand seat. *Key:* 1. Instrument panel with revolving Earth-globe; 2. Switch panel; 3. Television screen; 4. Television camera; 5. Porthole, incorporating 'Vzor' optical orientation device.

Soviet space activity. *Key:* 1. German collective, Khimki (1946); 2. German collective engine tests, Seliger Lake (1948); 3. U.S. Air Force radar surveillance, Samsun (Turkey); 4. Kapustin Yar cosmodrome (1947); 5. Tyuratam/Baikonur launch complex (1957); 6. Northern cosmodrome (1966); 7. Sounding rocket base,

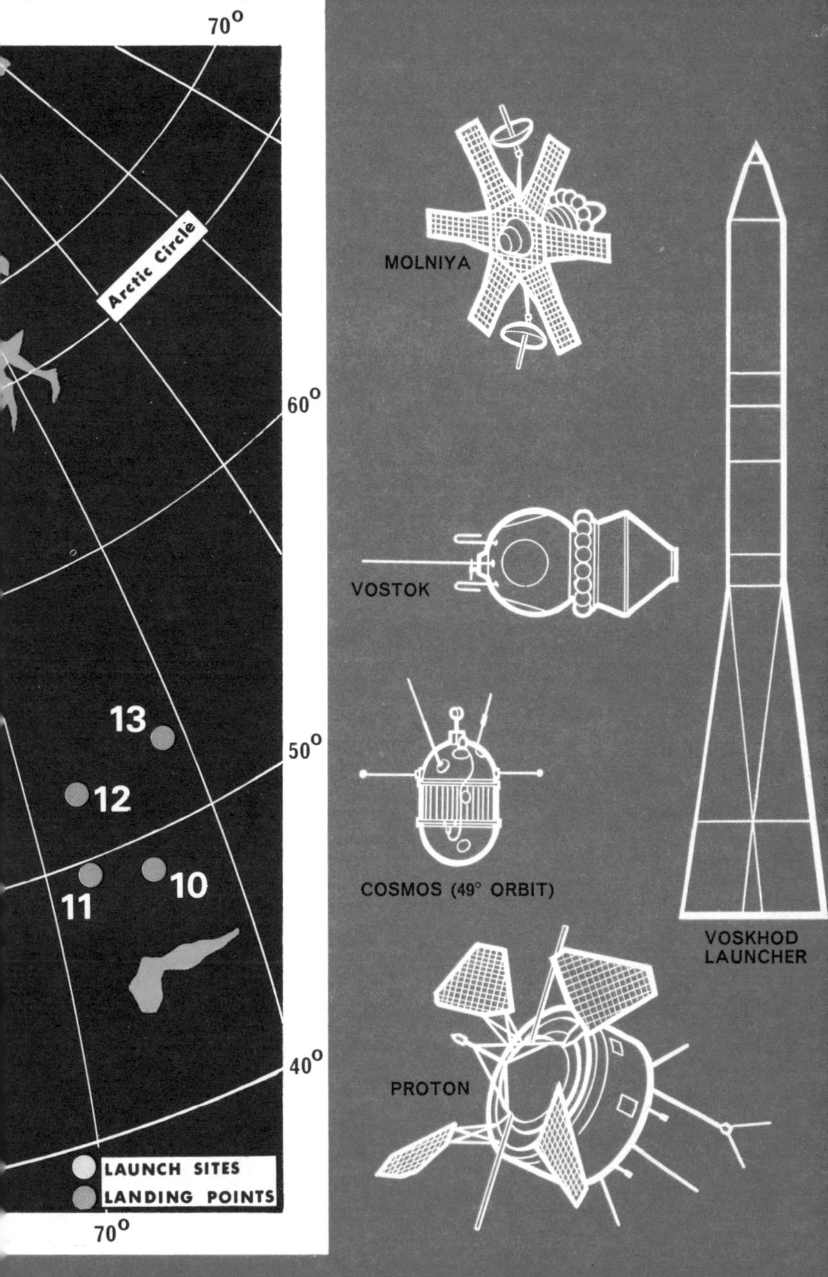

Franz Josef Land; following are landing places of early Soviet spacecraft: 8. Vostok 1; 9. Vostok 2; 10. Vostok 3; 11. Vostok 4; 12. Vostok 5; 13. Vostok 6; 14. Voskhod 1; 15. Voskhod 2.

Astronaut Walter Schirra's **Mercury capsule Sigma 7** installed on the Atlas-D launch vehicle at Cape Canaveral.

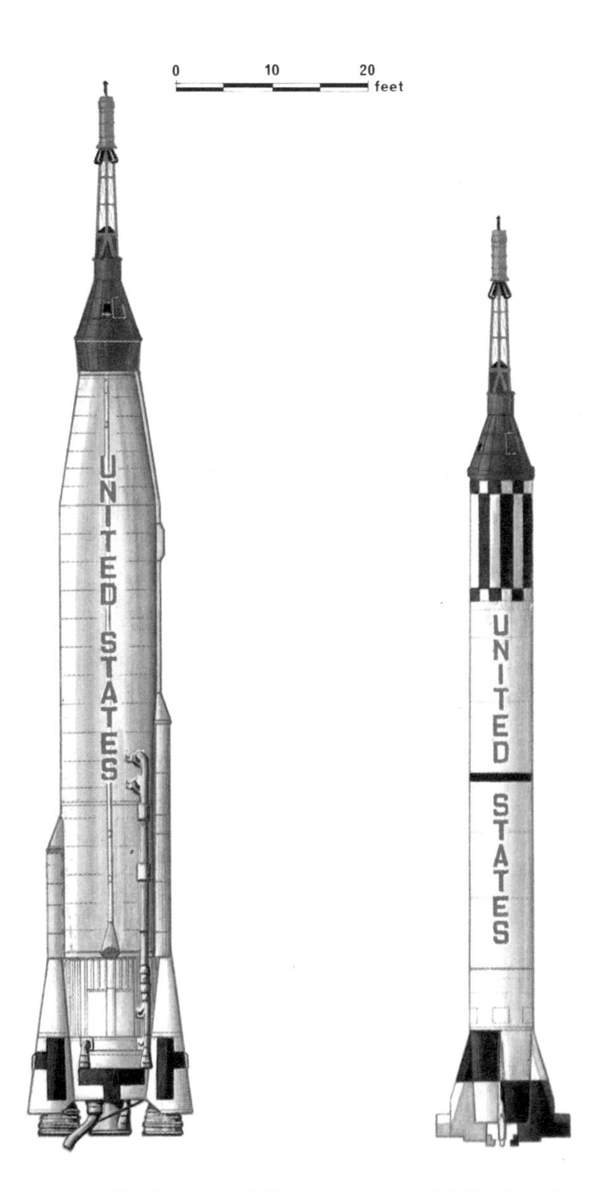

(Left) **Mercury-Atlas.** Length, including escape tower, 94 ft 4 in.; tank diameter 10 ft; width over boost fairing 16 ft. Launch weight 260,000 lb; launch thrust 367,000 lb. *(Right)* **Mercury-Redstone.** Length, including escape tower, 83 ft; diameter 5 ft 10 in. Launch weight 66,000 lb; launch thrust 78,000 lb.

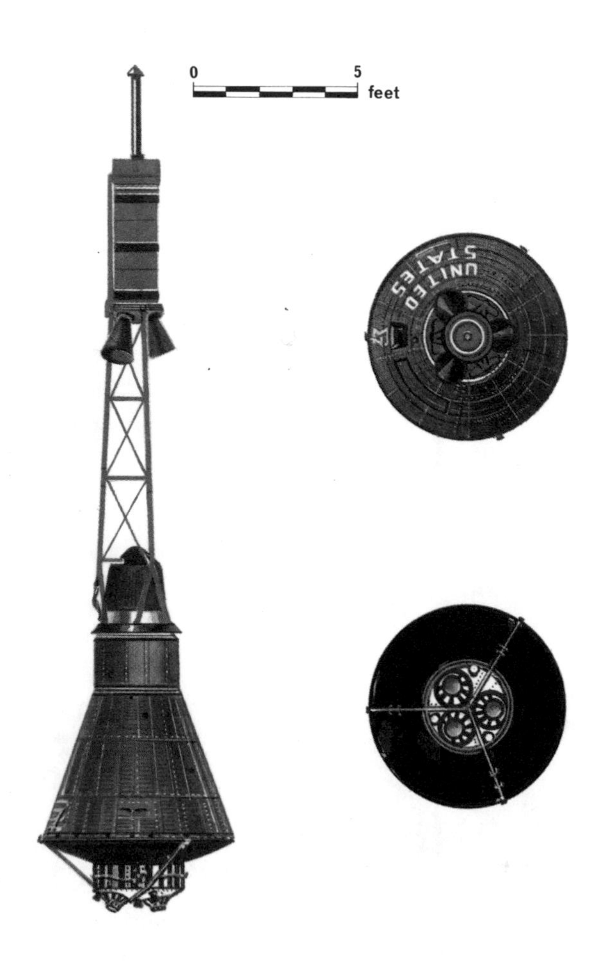

McDonnell Mercury. Length, including retro-pack, escape rocket and aerodynamic 'spike', 25 ft 11 in.; heat shield diameter 74·5 in. Typical weights (MA-6): at launch 4,265 lb; in orbit 2,987 lb; on recoverv 2,422 lb. Launch vehicle: modified Atlas-D.

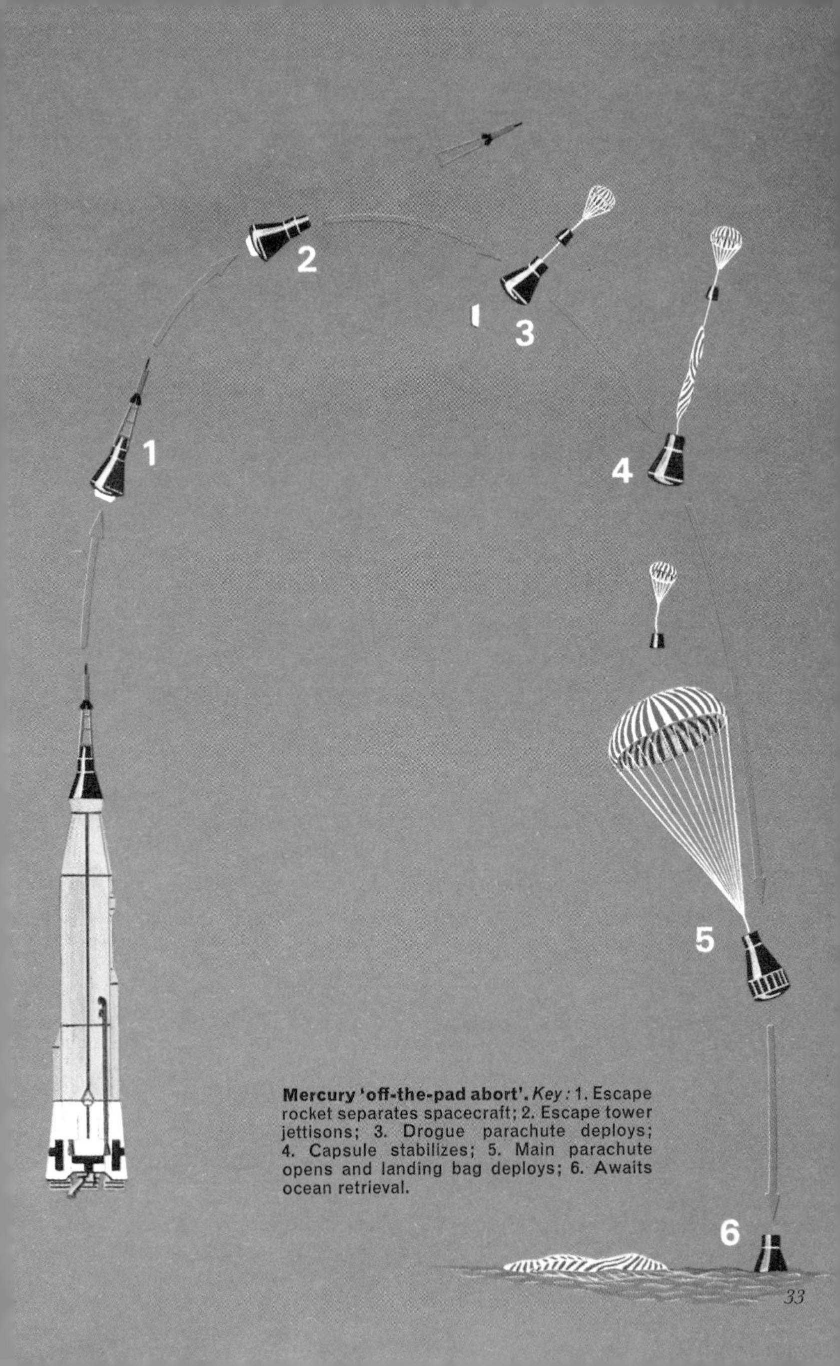

Mercury 'off-the-pad abort'. *Key:* 1. Escape rocket separates spacecraft; 2. Escape tower jettisons; 3. Drogue parachute deploys; 4. Capsule stabilizes; 5. Main parachute opens and landing bag deploys; 6. Awaits ocean retrieval.

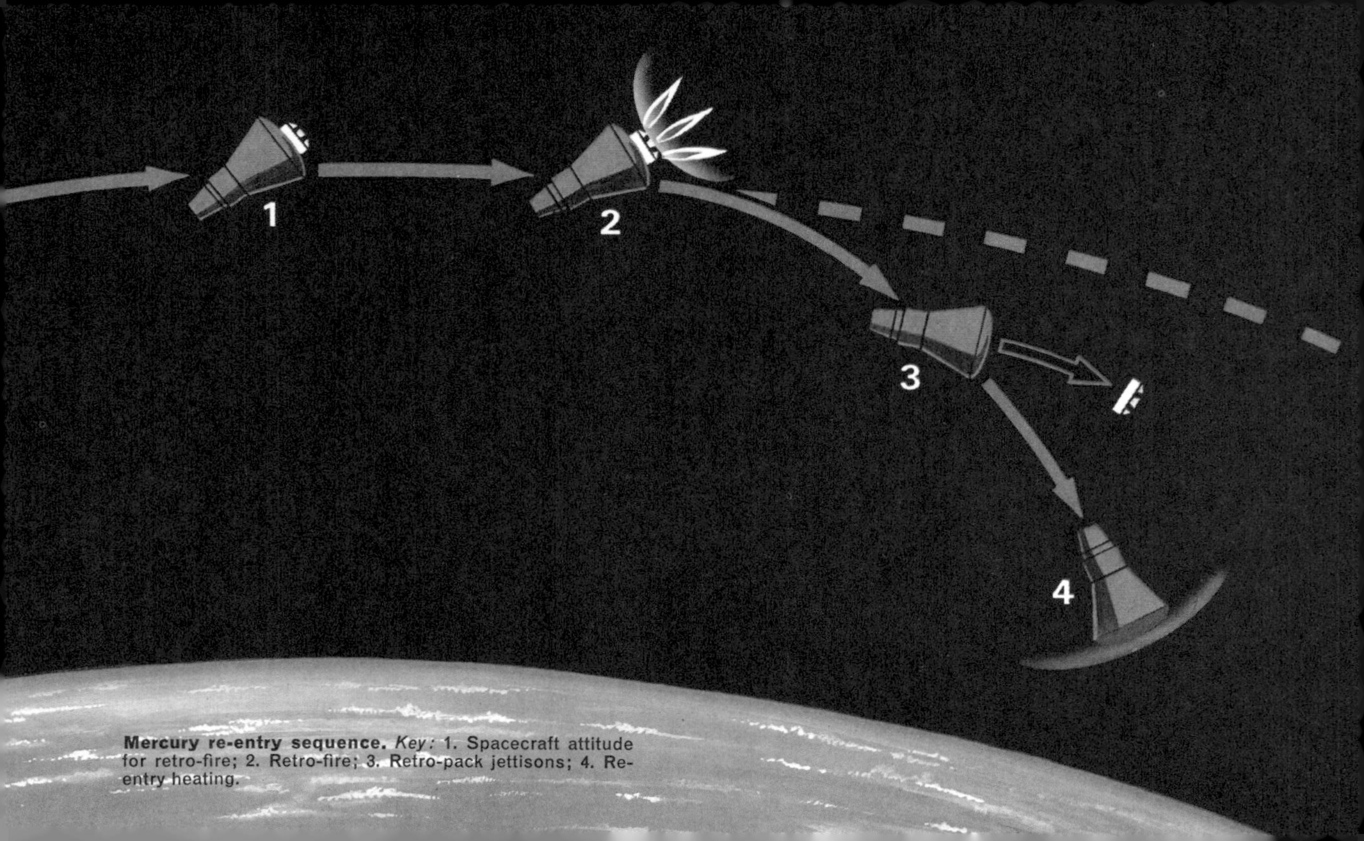

Mercury re-entry sequence. *Key:* 1. Spacecraft attitude for retro-fire; 2. Retro-fire; 3. Retro-pack jettisons; 4. Re-entry heating.

Mercury recovery sequence. *Key:* 1. Re-entry attitude; 2. Drogue deploys; 3. Main parachute deploys/parachute 'can' jettisons; 4. Main parachute reefed; 5. Landing parachute fully open and landing bag deployed; 6. Awaits ocean retrieval.

Gemini 6 with astronauts Walter Schirra and Thomas Stafford heads for an appointment in orbit with Gemini 7 on 15 December 1965. Within hours the two craft were to rendezvous at 17,500 m.p.h., keeping station only a few feet apart.

(*Left*) **Gemini-Titan 2.** Length 33·2 m. (109 ft); diameter 3·05 m. (10 ft). Launch weight 154,220 kg. (340,000 lb). First stage thrust 195,040 kg. (430,000 lb); second stage thrust 45,359 kg (100,000 lb). (*Right*) **MOL-Titan 3M** (cancelled 1969). Length 47·2 m. (155 ft); Titan core diameter 3·05 m. (10 ft); seven-segment lateral solid boosters 33·5 m. (110 ft) long × 3·05 m. (10 ft) diameter. Designed to launch Manned Orbiting Laboratory, also cancelled 1969, for military reconnaissance and surveillance by two-man crew using advanced optical and electronic sensors. **MOL.** Length c/w Gemini-type spacecraft 16·4 m. (54 ft), less spacecraft 12·5 m. (41 ft), diameter 3·05 m. (10 ft).

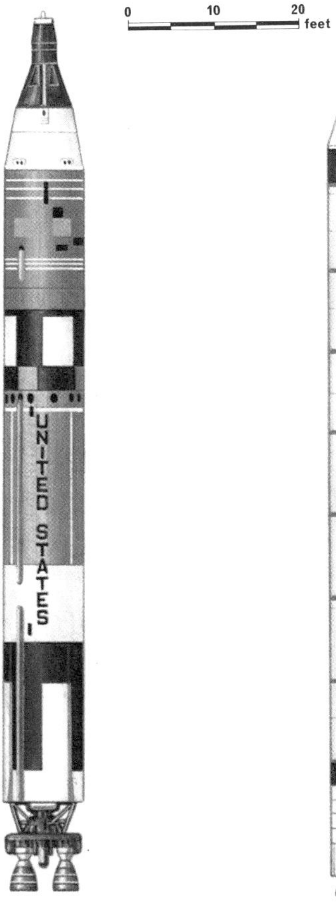

0 10 20 feet

UNITED STATES

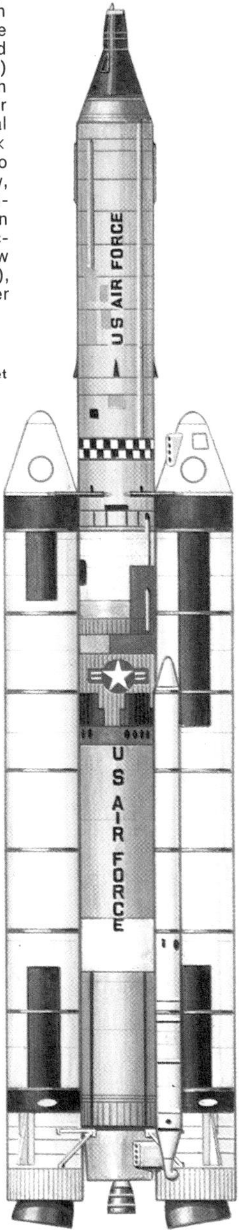

US AIR FORCE

US AIR FORCE

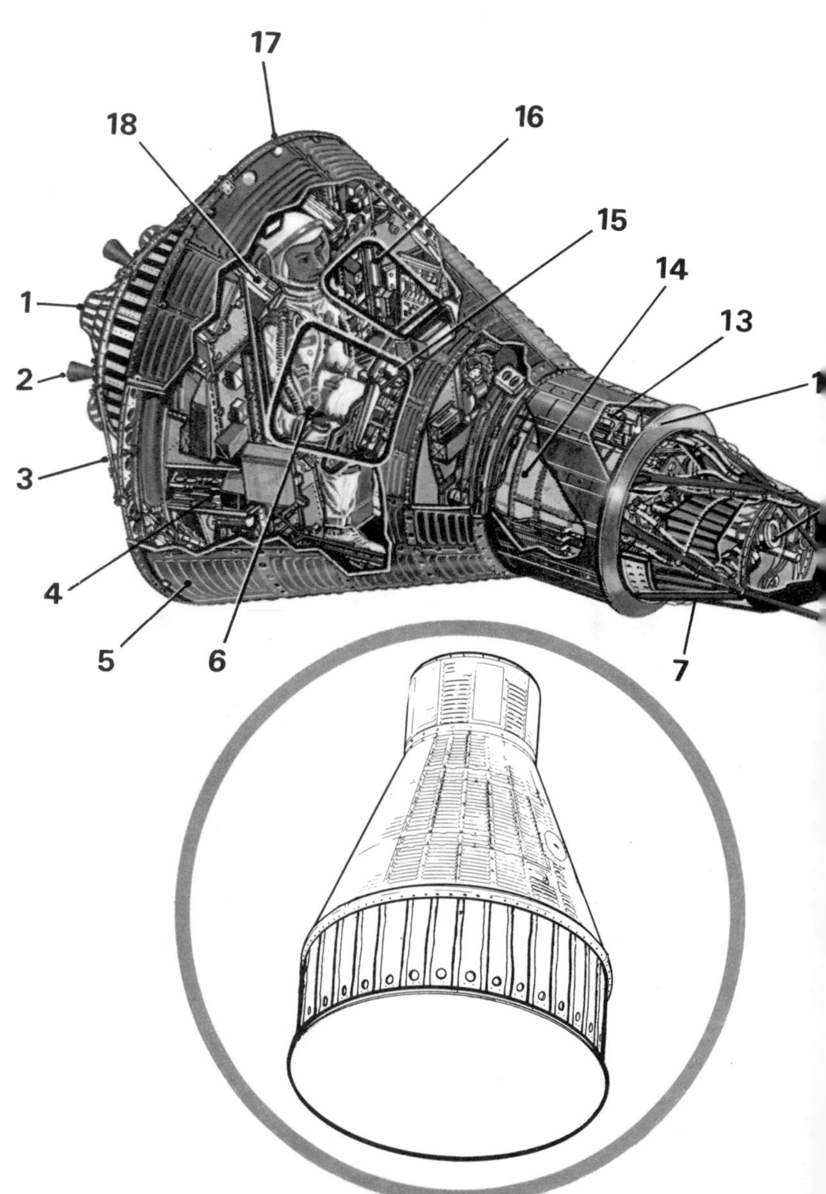
17
18
16
15
14
13
1
2
3
4
5
6
7

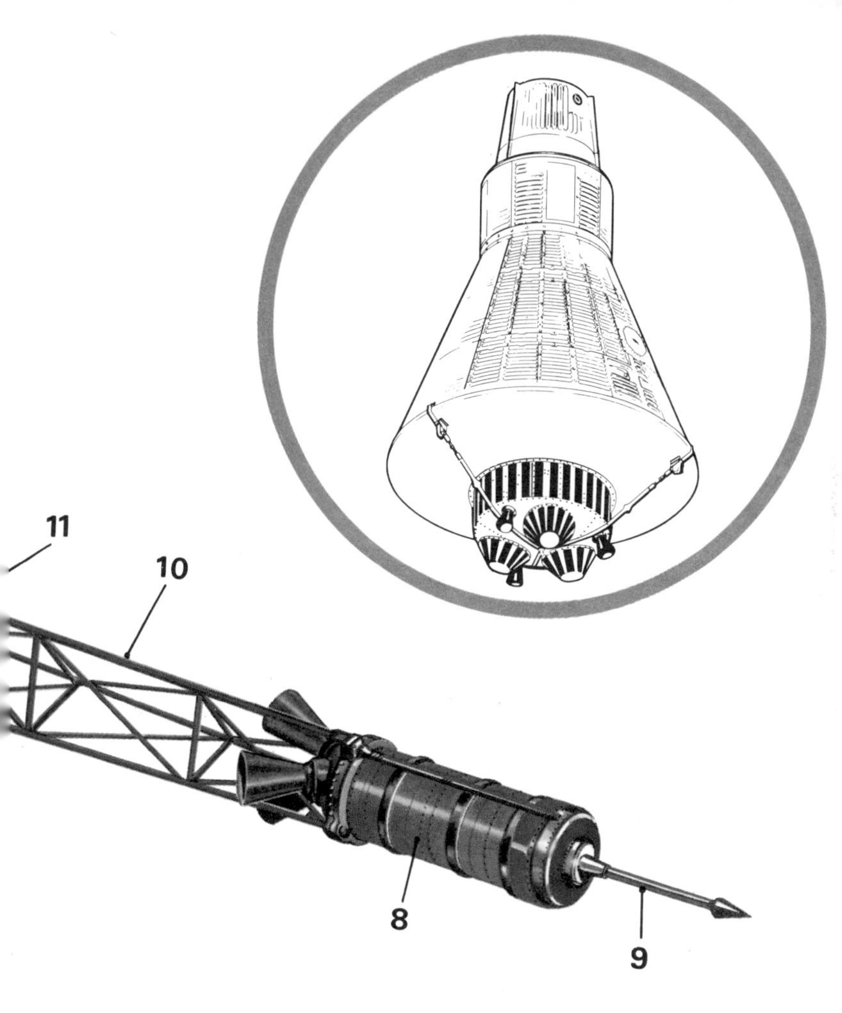

McDonnell Mercury. *Key:* 1. Retro-pack; 2. Separation rockets; 3. Retro-rocket retaining straps; 4. Attitude thrusters; 5. Skin shingles; 6. Hatch; 7. Drogue parachute stowage; 8. Emergency escape rocket; 9. Aerodynamic spike; 10. Escape tower; 11. Horizon sensors; 12. Aerodynamic fairing; 13. Attitude thrusters; 14. Main and reserve landing parachutes; 15. Instrument panels; 16. Window; 17. Heat shield; 18. Astronaut's couch.

(Inset, right) shows the ablative heat shield and retro-pack which also contained spacecraft separation rockets;

(inset, left) shows the heat shield dropped to extend the pneumatic landing bag.

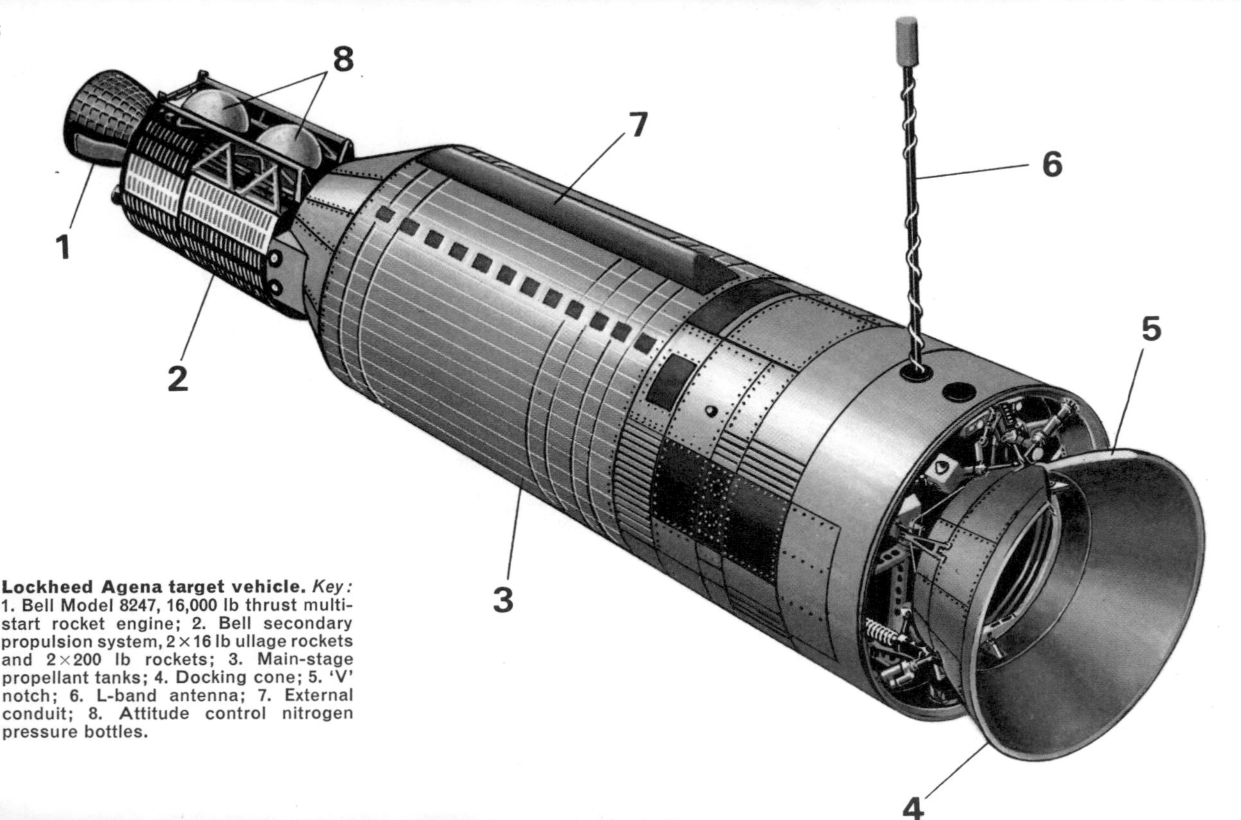

Lockheed Agena target vehicle. *Key:* 1. Bell Model 8247, 16,000 lb thrust multi-start rocket engine; 2. Bell secondary propulsion system, 2×16 lb ullage rockets and 2×200 lb rockets; 3. Main-stage propellant tanks; 4. Docking cone; 5. 'V' notch; 6. L-band antenna; 7. External conduit; 8. Attitude control nitrogen pressure bottles.

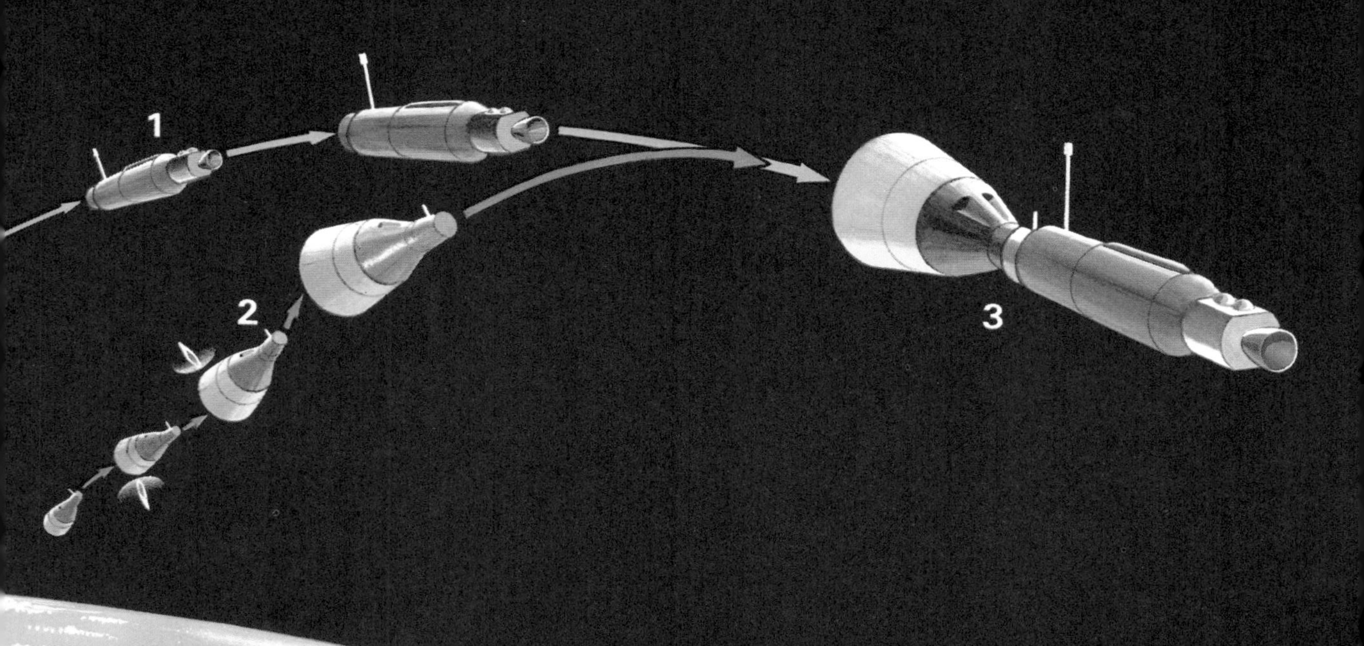

Gemini rendezvous and docking.
Key: 1. Agena target vehicle; 2. Gemini radar-controlled manœuvre; 3. Gemini docking under manual sighting and control.

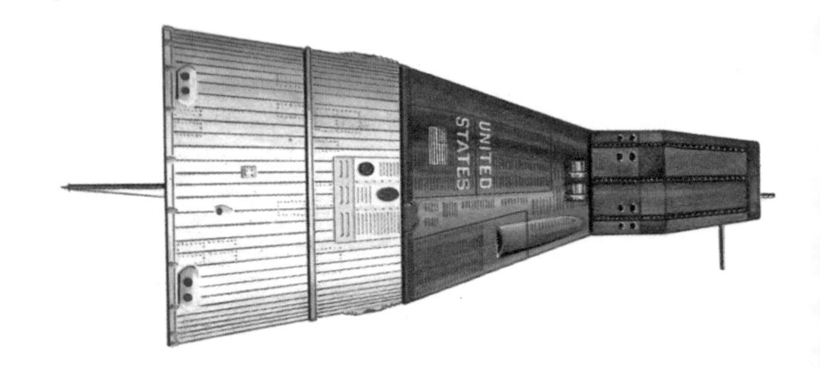

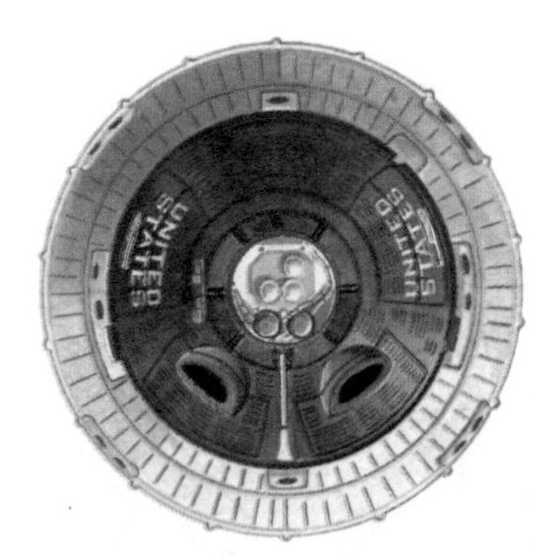

McDonnell Gemini. Length complete with retrograde and equipment modules 18 ft 5 in.; maximum diameter 10 ft; small end diameter 39 in.; length retrograde and equipment modules 7 ft 6 in. Heat shield diameter 90 in. Weight: 7,100 to 8,350 lb. Launch vehicle: modified Titan 2.

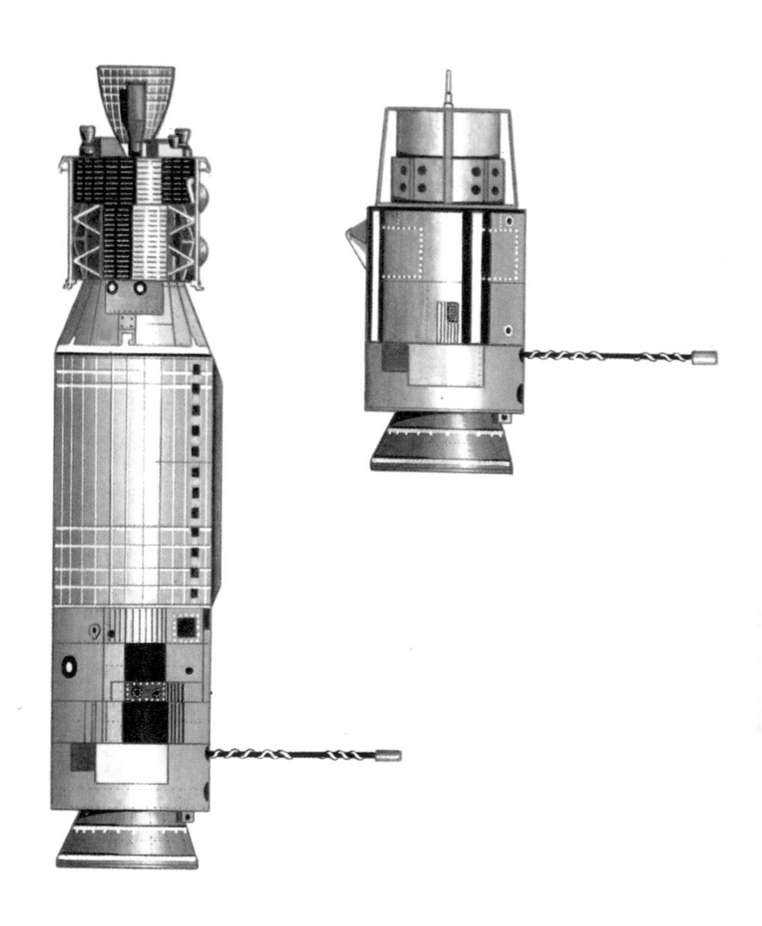

(Left) **Lockheed Agena rendezvous target.** Length 25 ft 7 in.; diameter 5 ft. Weight, fuelled, in orbit 7,000 lb. Main engine thrust 16,000 lb. *(Right)* **McDonnell A.T.D.A.** Length 10 ft 11 in.; diameter 5 ft. Typical weights: at launch 2,400 lb; in orbit 1,700 lb. No independent propulsion. Launch vehicle, both targets, modified Atlas-D (overall height with Agena 104 ft; with A.T.D.A. 95 ft 6 in.).

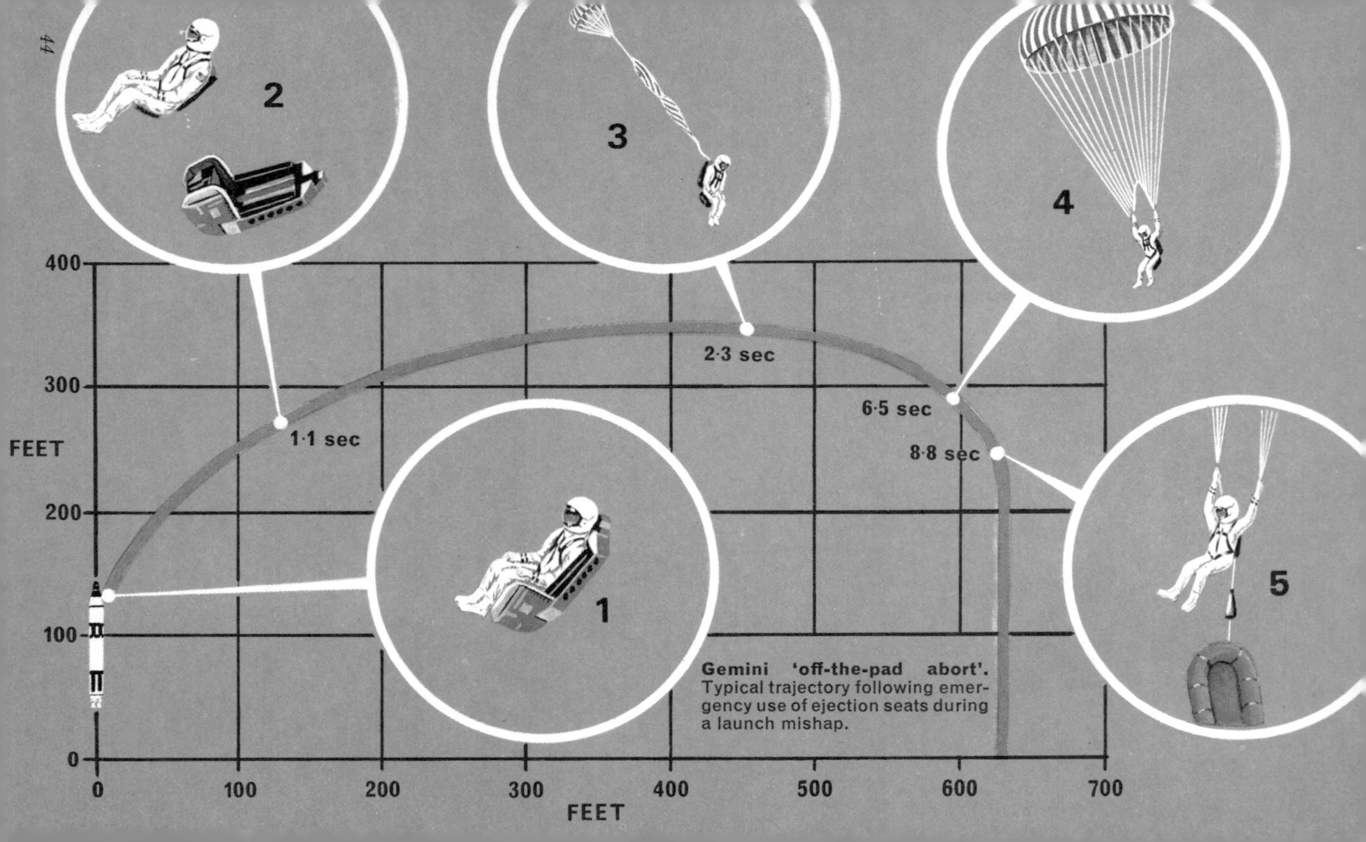

Gemini 'off-the-pad abort'. Typical trajectory following emergency use of ejection seats during a launch mishap.

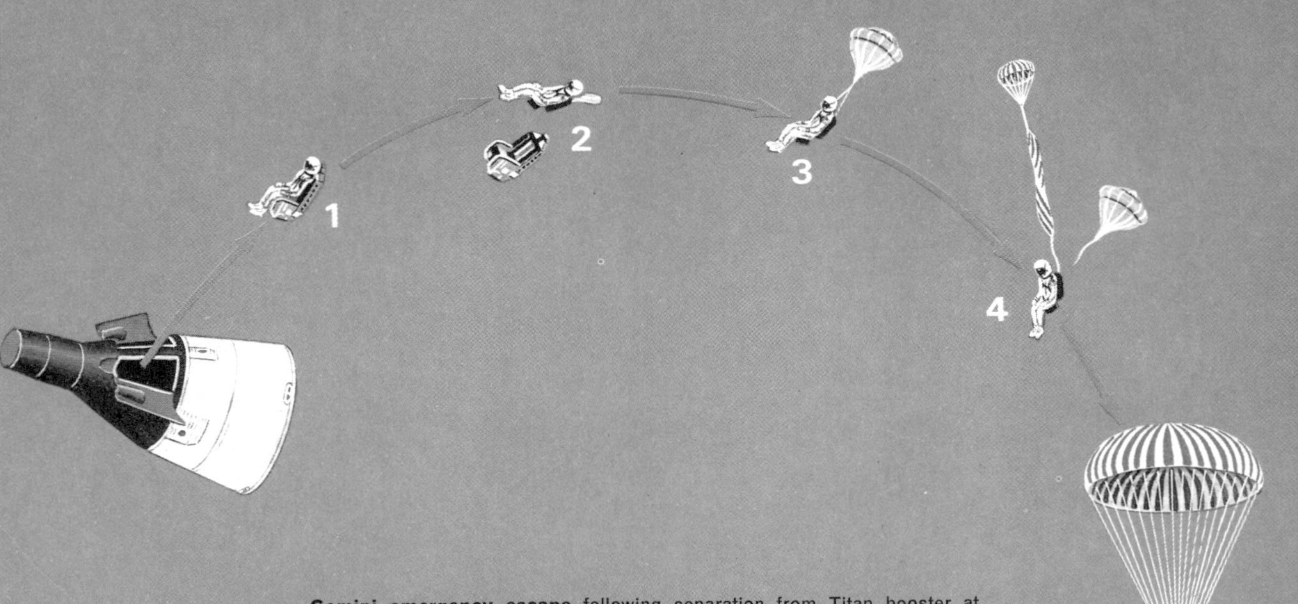

Gemini emergency escape following separation from Titan booster at higher altitude. *Key:* 1. Seat ejection; 2. Astronaut free of seat; 3. 'Ballute' stabilizer inflates; 4. Drogue deploys/ballute jettisons; 5. Landing parachute fully inflated.

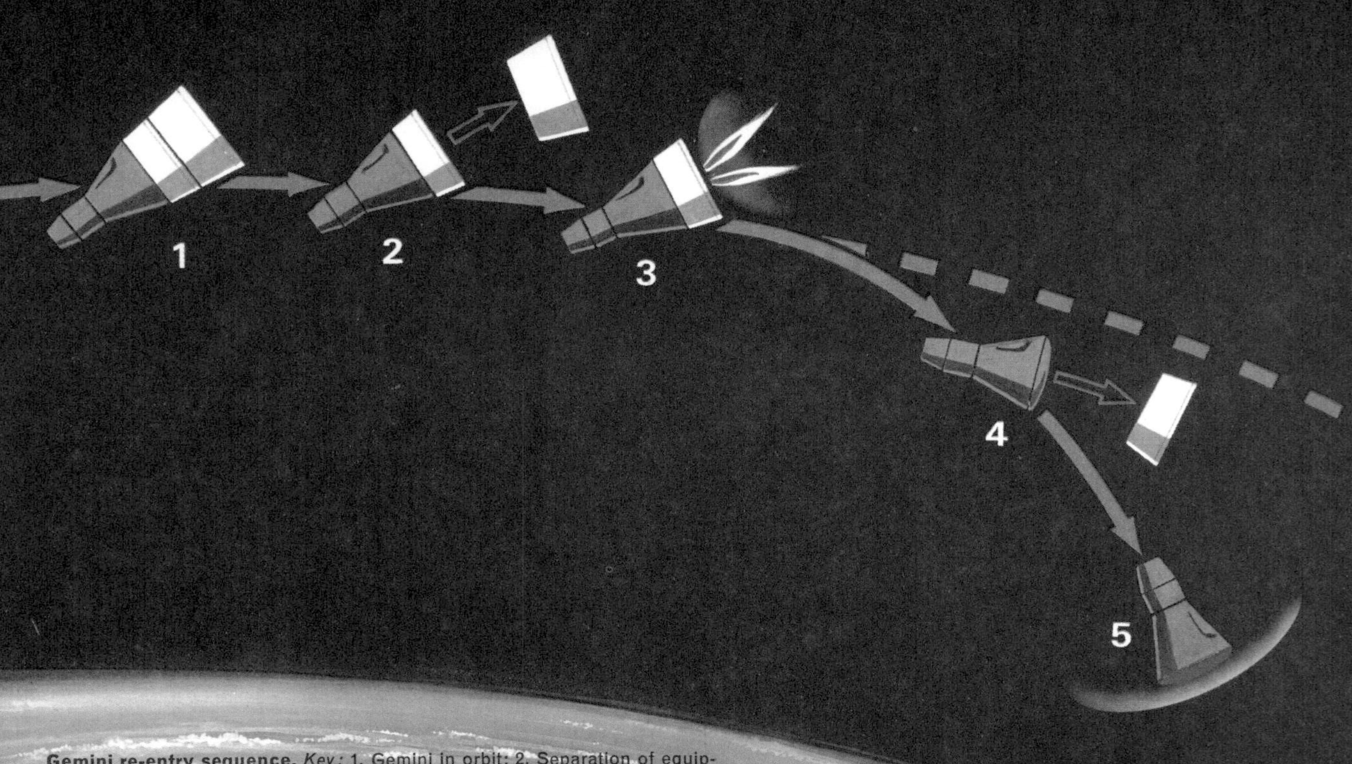

Gemini re-entry sequence. *Key:* 1. Gemini in orbit; 2. Separation of equipment module; 3. Retro-fire; 4. Separation of retro-module; 5. Re-entry heating.

Gemini recovery sequence. *Key:* 1. Drogue parachute deploys; 2. Parachute 'can' separates; 3. Main parachute deploys; 4. Main parachute reefed; 5. Spacecraft assumes landing attitude; 6. Awaiting ocean retrieval.

(Top) **Gemini 7** as seen from the window of Gemini 6 during the first space rendezvous; *(bottom)* instant before docking with an Agena target vehicle. The view from Gemini 10.

Gemini 11 astronauts look down on the Agena target vehicle with which their own craft is loosely connected by a 100 ft tether. They were orbiting over Lower California near La Paz.

(Top) Edward White becomes the first American to leave a spacecraft in orbit on 3 June 1965. *(Bottom)* Richard Gordon straddles the cylindrical neck of Gemini 11 to fix a Dacron tether between his craft and the docked Agena.

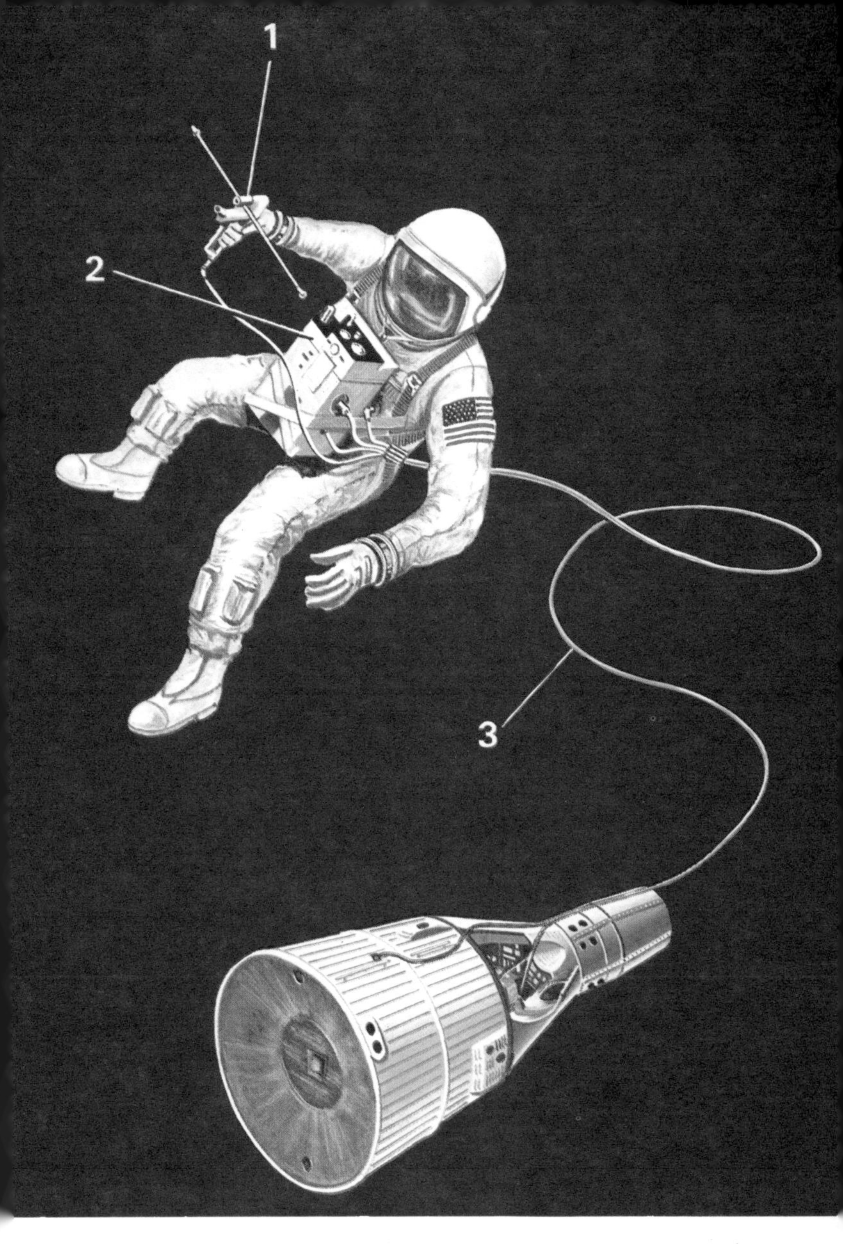

Extra-vehicular astronaut. Helmet visor thinly coated with gold protects the eyes from intense sunlight in space but allows clear visibility. *Key:* 1. Hand-held manœuvring unit; 2. Extra-vehicular life support system; 3. Astronaut umbilical tether.

North-Western Australia as seen from Gemini 11 from 850 miles. At extreme left is the city of Perth; *top left*, Indian Ocean; *left* of antenna tip, North West Cape; *top*, Timor Sea and *extreme top right*, Indonesia. In the *foreground*, near antenna stem, is the Great Sandy Desert.

Red Sea and Gulf of Aden from Gemini 11. Antenna points to Red Sea, Saudi Arabia is above; Ethiopia below and Indian Ocean at extreme right. It is instructive to compare these photographs with appropriate maps in an atlas.

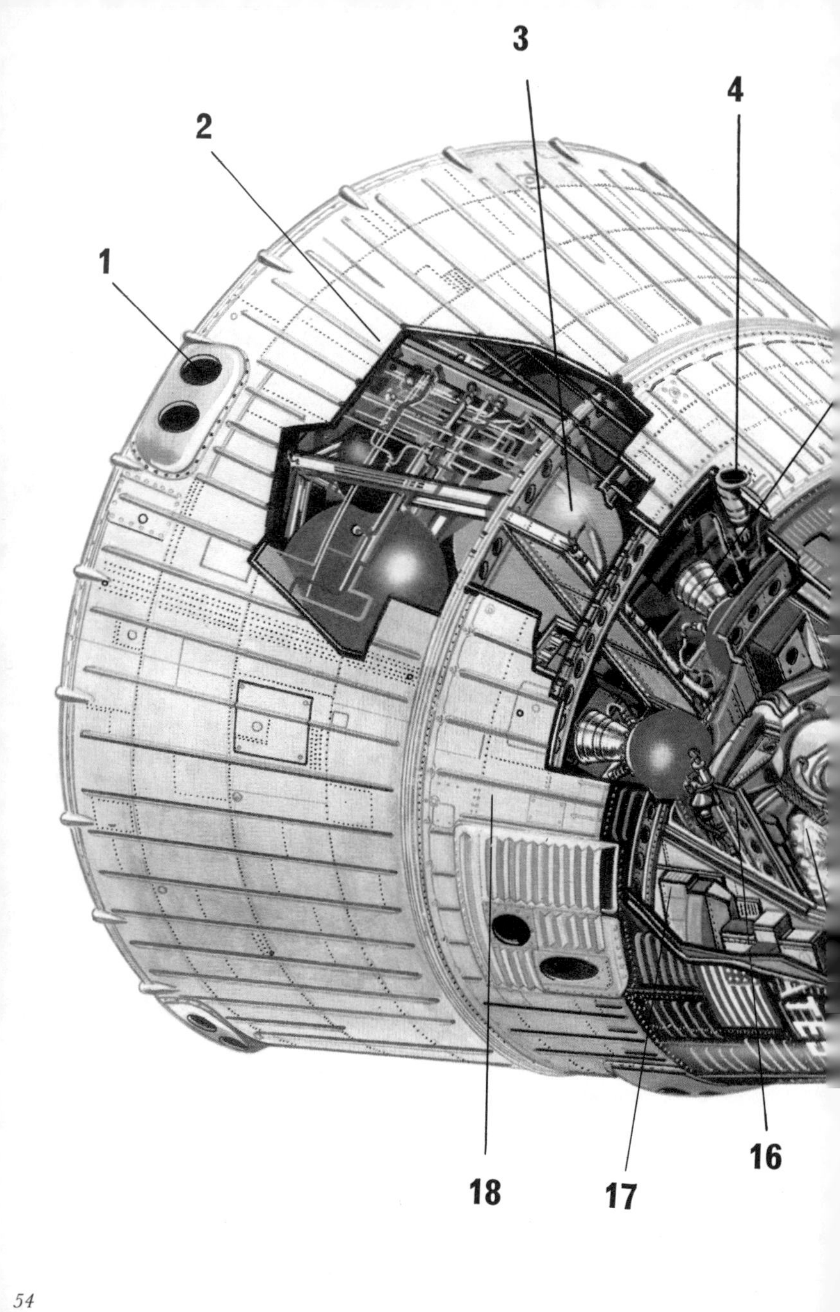
1
2
3
4
16
17
18

McDonnell Gemini. *Key:* 1. Attitude control thrusters; 2. Equipment module; 3. Fuel, oxidant and pressure tanks; 4. Manœuvre thrusters; 5. Retro-rockets; 6. Command pilot; 7. Window; 8. Re-entry attitude thrusters; 9. Landing parachute stowage; 10. Docking bar; 11. Rendezvous radar; 12. Drogue parachute stowage; 13. Re-entry module; 14. Instrument panels; 15. Pilot/EVA astronaut; 16. Ejection seat; 17. Electrical equipment; 18. Retro-module.

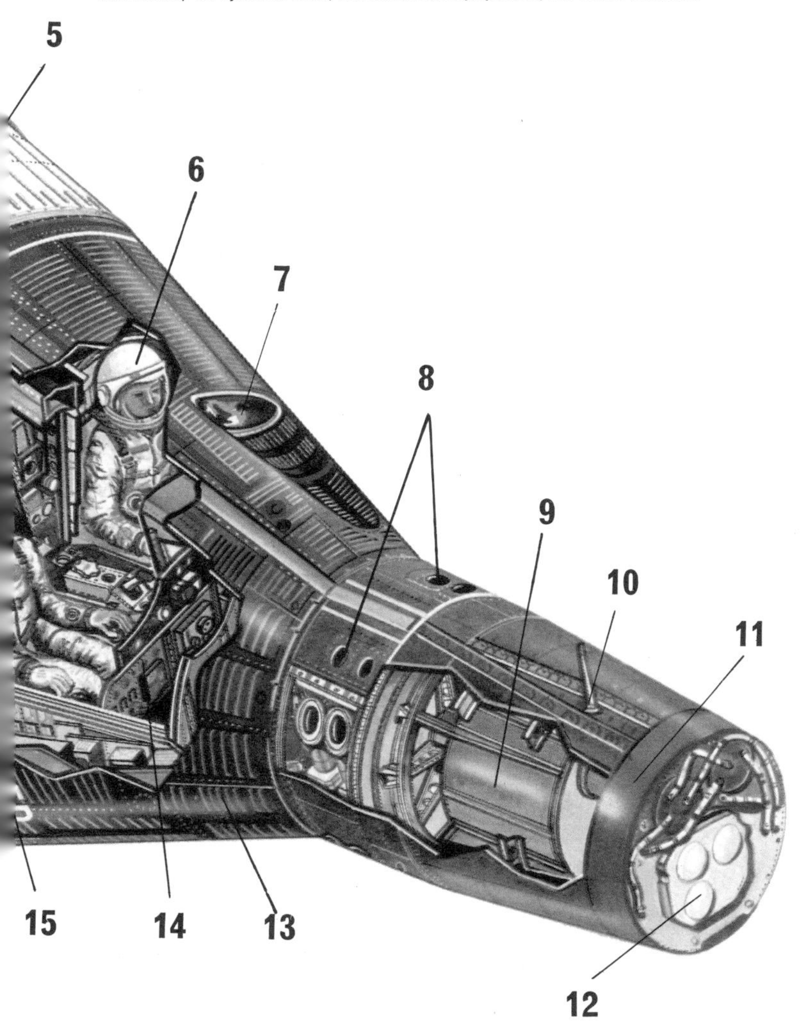

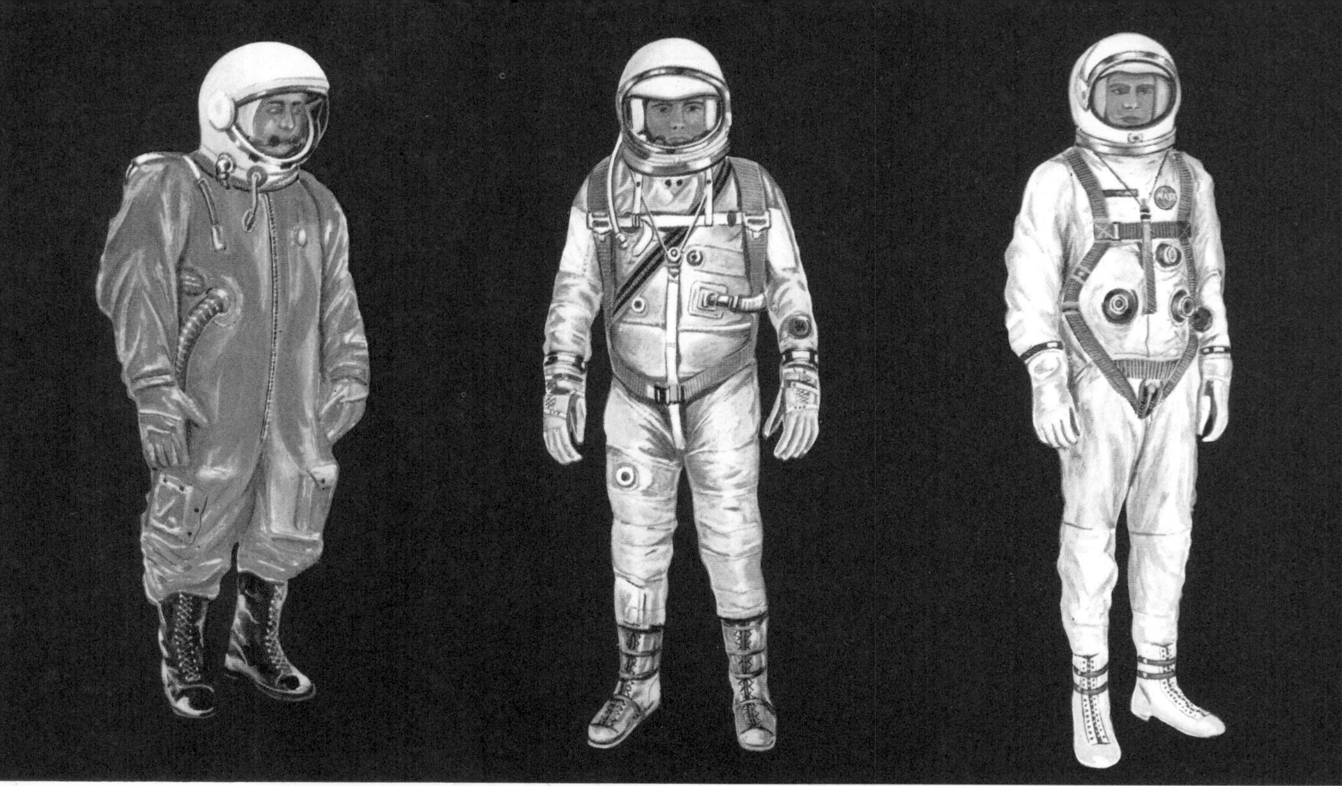

Early spacesuits: *(left to right)* Vostok; Mercury and Gemini. In the case of Vostok the actual pressure garment was hidden by a loose-fitting coverall.

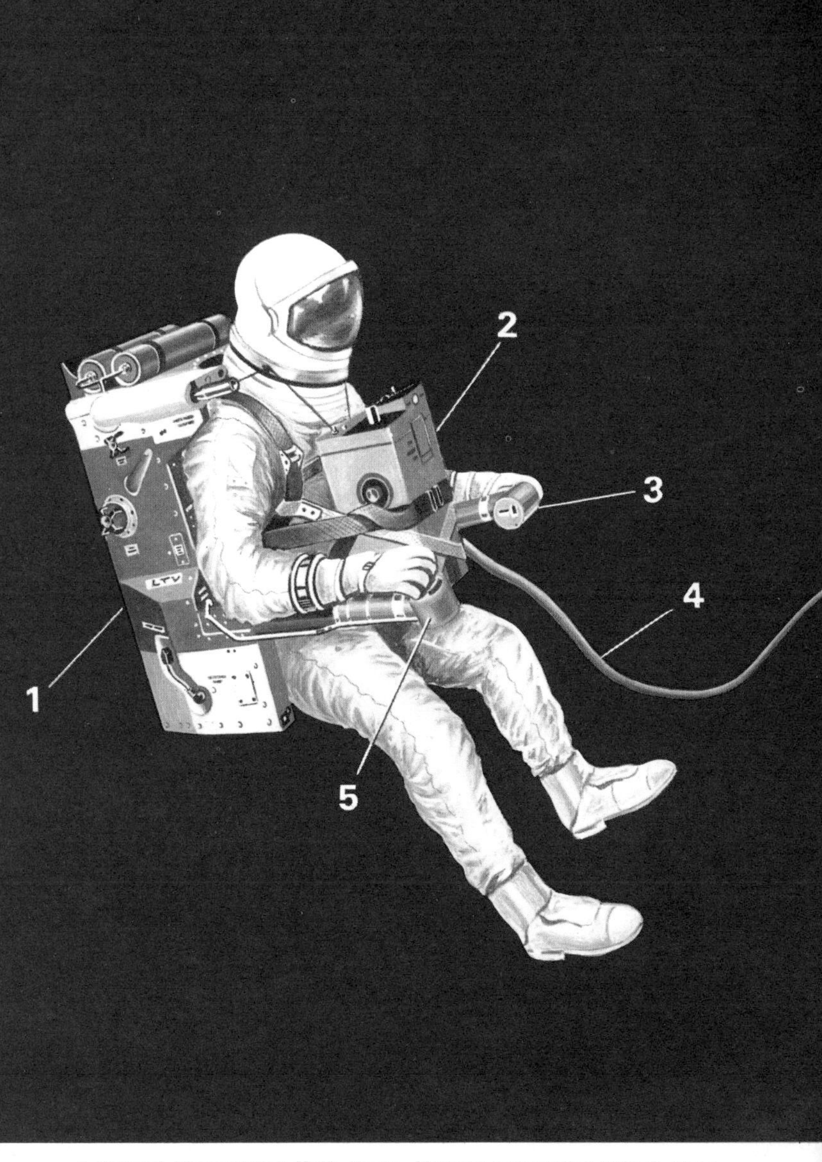

Astronaut Manœuvring Unit. *Key:* 1. Modular manœuvring unit; 2. Emergency life support system; 3. Arm-rest and hand control; 4. Tether; 5. Arm-rest and hand control.

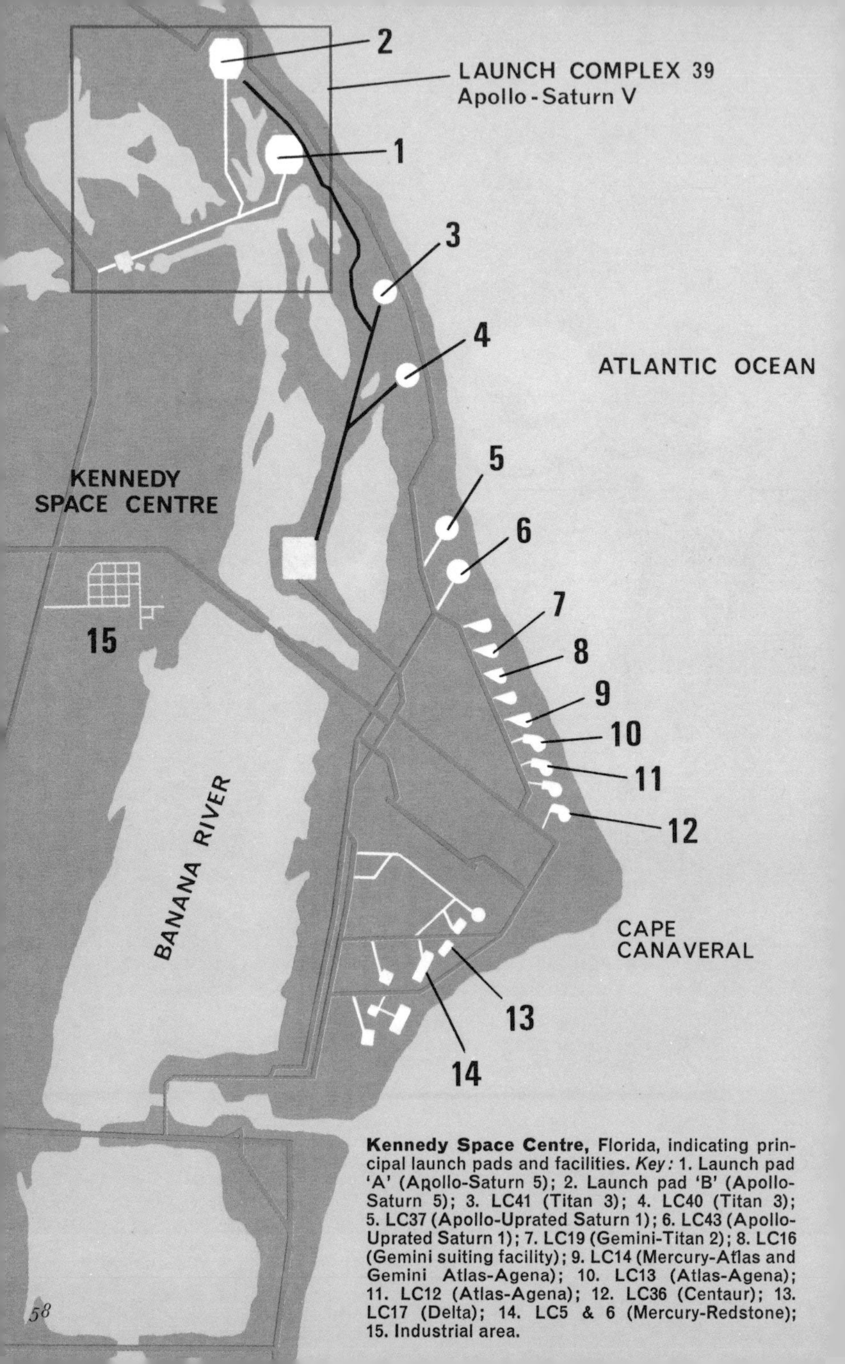

Kennedy Space Centre, Florida, indicating principal launch pads and facilities. *Key:* 1. Launch pad 'A' (Apollo-Saturn 5); 2. Launch pad 'B' (Apollo-Saturn 5); 3. LC41 (Titan 3); 4. LC40 (Titan 3); 5. LC37 (Apollo-Uprated Saturn 1); 6. LC43 (Apollo-Uprated Saturn 1); 7. LC19 (Gemini-Titan 2); 8. LC16 (Gemini suiting facility); 9. LC14 (Mercury-Atlas and Gemini Atlas-Agena); 10. LC13 (Atlas-Agena); 11. LC12 (Atlas-Agena); 12. LC36 (Centaur); 13. LC17 (Delta); 14. LC5 & 6 (Mercury-Redstone); 15. Industrial area.

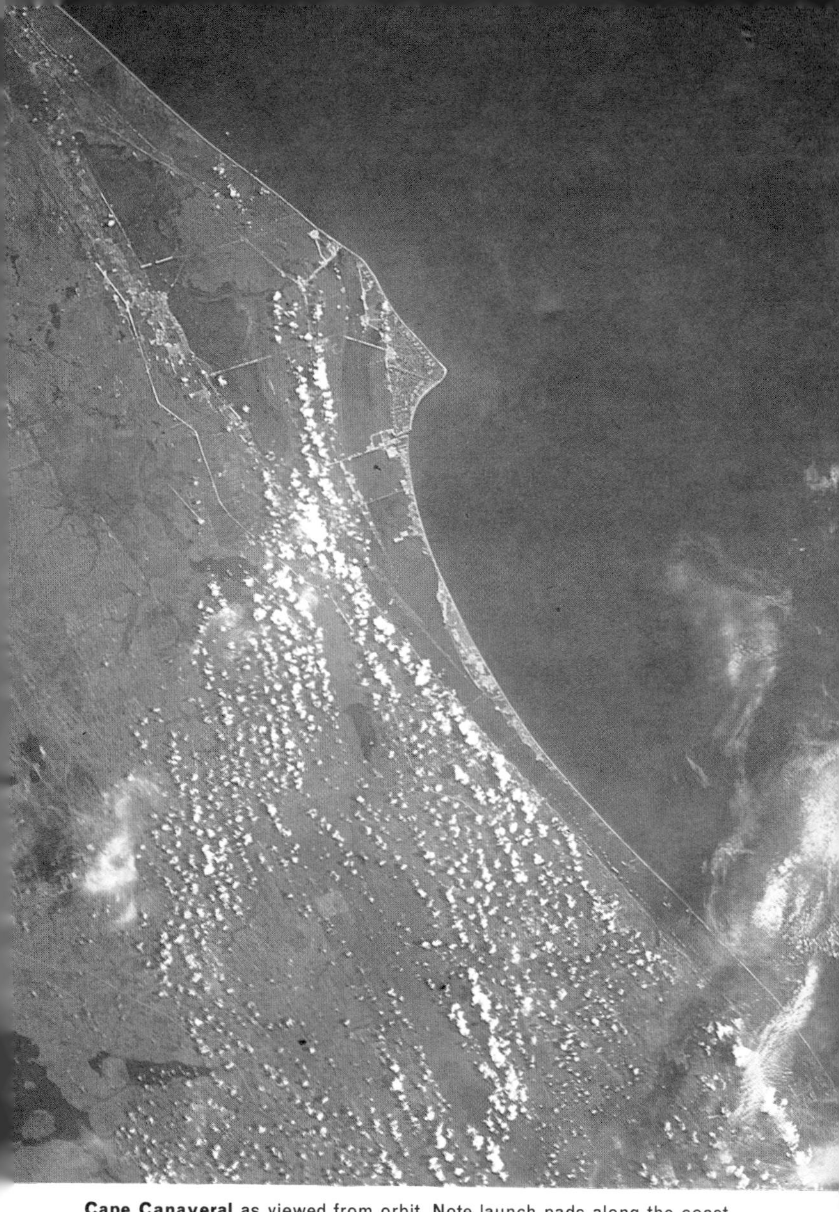

Cape Canaveral as viewed from orbit. Note launch pads along the coast, causeways over Banana River and Apollo launch pads on Merritt Island. The picture was taken by Gemini 5 astronaut Charles Conrad from 130 miles altitude.

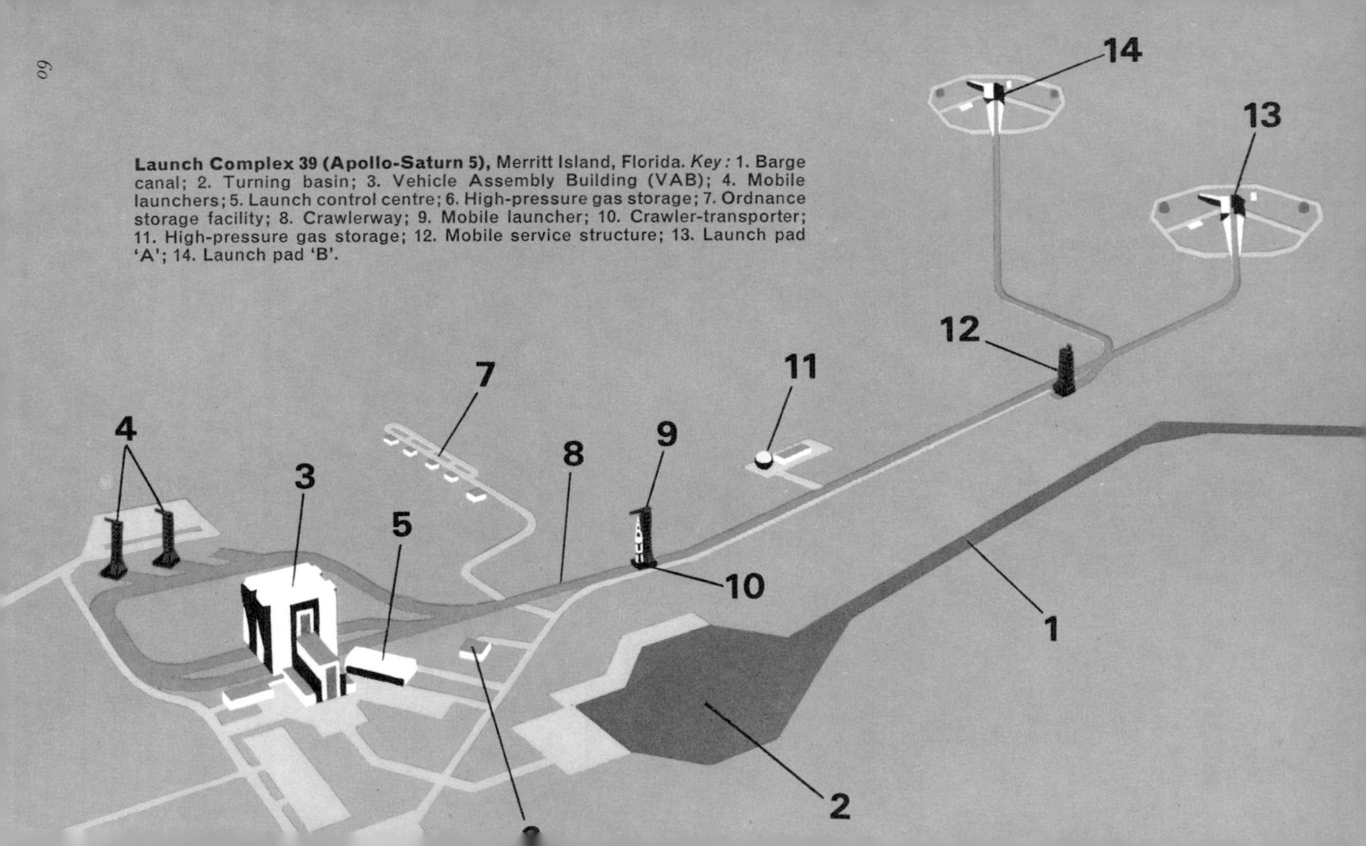

Launch Complex 39 (Apollo-Saturn 5), Merritt Island, Florida. *Key:* 1. Barge canal; 2. Turning basin; 3. Vehicle Assembly Building (VAB); 4. Mobile launchers; 5. Launch control centre; 6. High-pressure gas storage; 7. Ordnance storage facility; 8. Crawlerway; 9. Mobile launcher; 10. Crawler-transporter; 11. High-pressure gas storage; 12. Mobile service structure; 13. Launch pad 'A'; 14. Launch pad 'B'.

The Saturn 5 launch complex on Merritt Island, Florida, with the Vehicle Assembly Building in the foreground. In the far distance are launch pads 'A' and 'B' for project Apollo.

0 50
feet

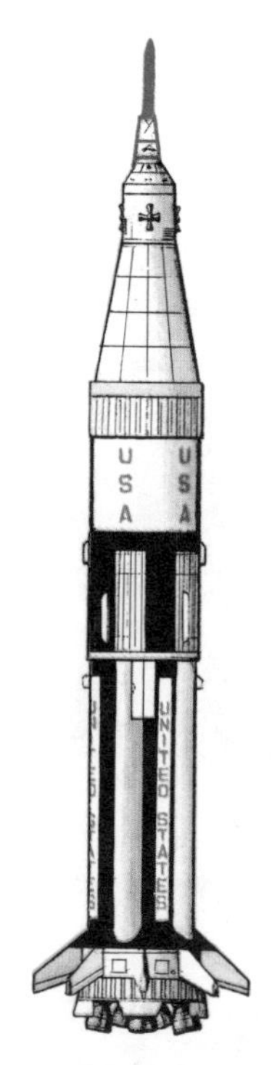

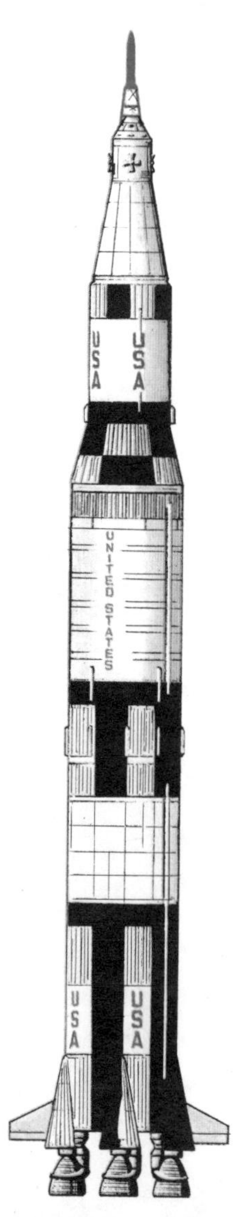

(Left) **Apollo-Saturn 1B.** Length including spacecraft escape tower 224 ft; maximum diameter 21 ft 8 in. Launch weight 1,300,000 lb; launch thrust 1,600,000 lb. *(Right)* **Apollo-Saturn 5.** Length including spacecraft escape tower 364 ft 6 in.; maximum diameter 33 ft. Launch weight 6,000,000+ lb; launch thrust 7,500,000+ lb (see page 189).

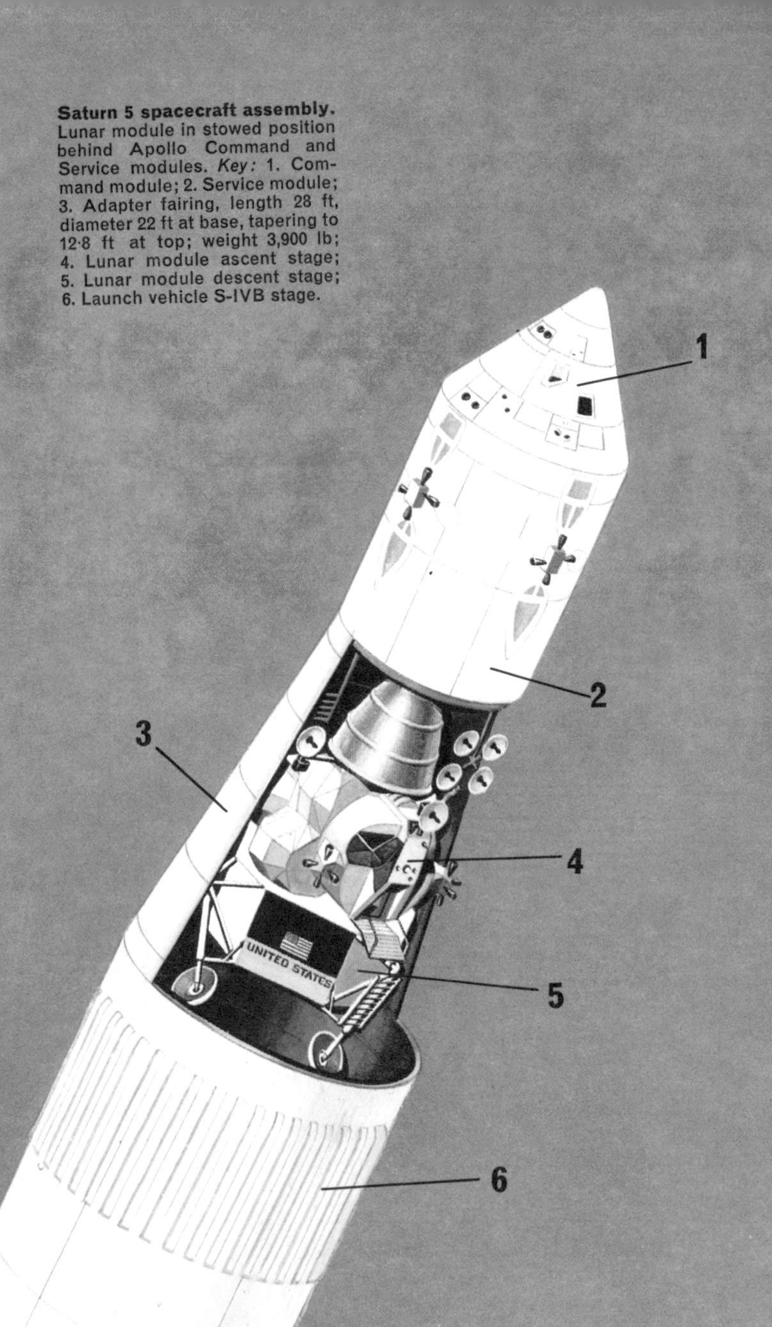

Saturn 5 spacecraft assembly. Lunar module in stowed position behind Apollo Command and Service modules. *Key:* 1. Command module; 2. Service module; 3. Adapter fairing, length 28 ft, diameter 22 ft at base, tapering to 12·8 ft at top; weight 3,900 lb; 4. Lunar module ascent stage; 5. Lunar module descent stage; 6. Launch vehicle S-IVB stage.

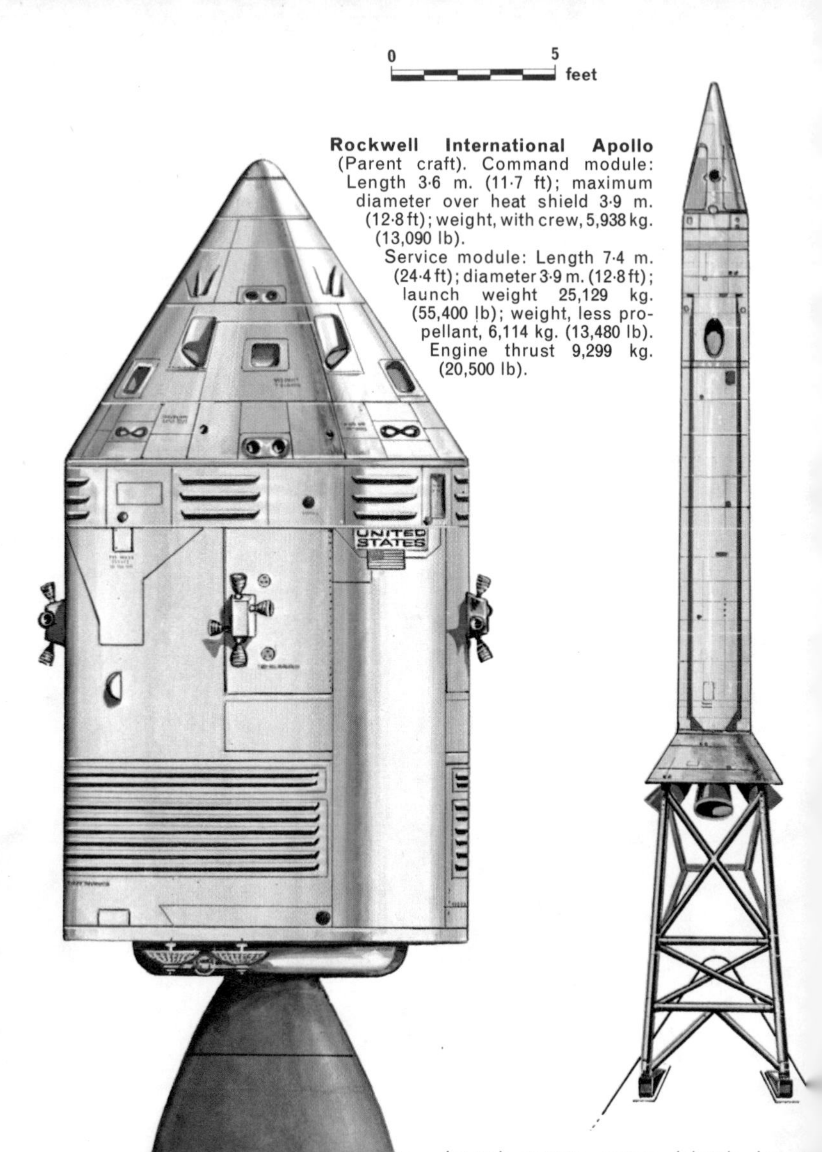

Rockwell International Apollo (Parent craft). Command module: Length 3·6 m. (11·7 ft); maximum diameter over heat shield 3·9 m. (12·8 ft); weight, with crew, 5,938 kg. (13,090 lb).

Service module: Length 7·4 m. (24·4 ft); diameter 3·9 m. (12·8 ft); launch weight 25,129 kg. (55,400 lb); weight, less propellant, 6,114 kg. (13,480 lb). Engine thrust 9,299 kg. (20,500 lb).

Launch escape system (*above*): Lengt overall c/w tower, 10·2 m. (33·4 ft); diamete 66 cm. (26 in.); weight 4,173 kg. (9,200 lb). Weights are not typical of every missior e.g. CSM launch weight for Apollo 11 wa 28,800 kg. (63,493 lb).

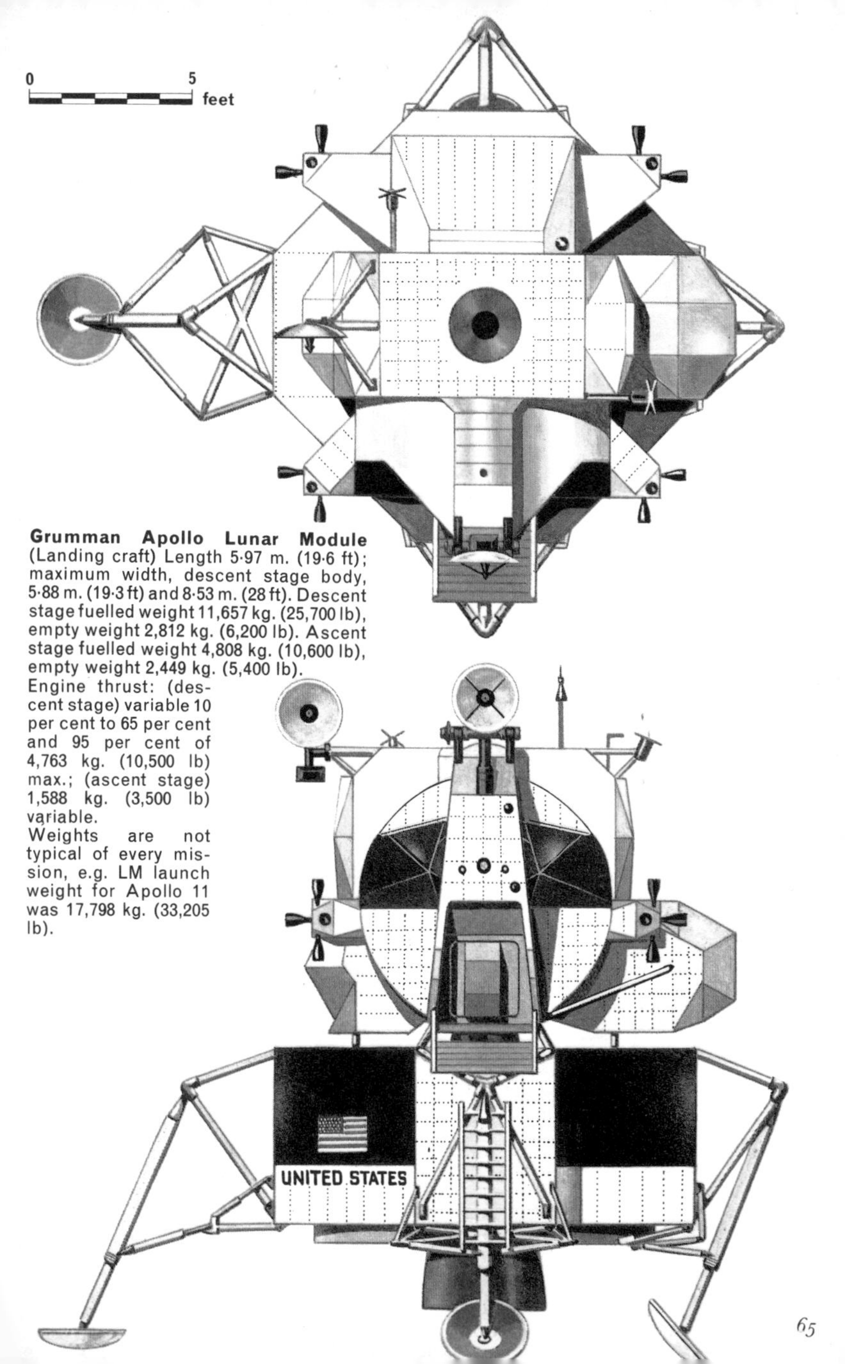

Grumman Apollo Lunar Module (Landing craft) Length 5·97 m. (19·6 ft); maximum width, descent stage body, 5·88 m. (19·3 ft) and 8·53 m. (28 ft). Descent stage fuelled weight 11,657 kg. (25,700 lb), empty weight 2,812 kg. (6,200 lb). Ascent stage fuelled weight 4,808 kg. (10,600 lb), empty weight 2,449 kg. (5,400 lb). Engine thrust: (descent stage) variable 10 per cent to 65 per cent and 95 per cent of 4,763 kg. (10,500 lb) max.; (ascent stage) 1,588 kg. (3,500 lb) variable.
Weights are not typical of every mission, e.g. LM launch weight for Apollo 11 was 17,798 kg. (33,205 lb).

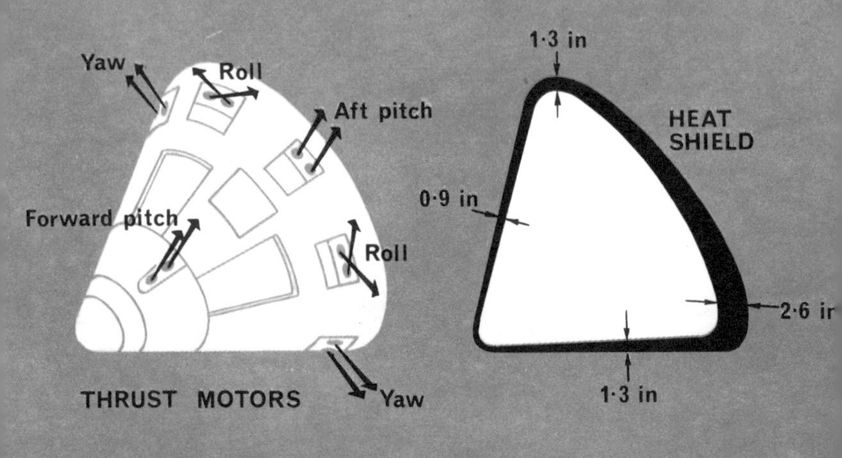

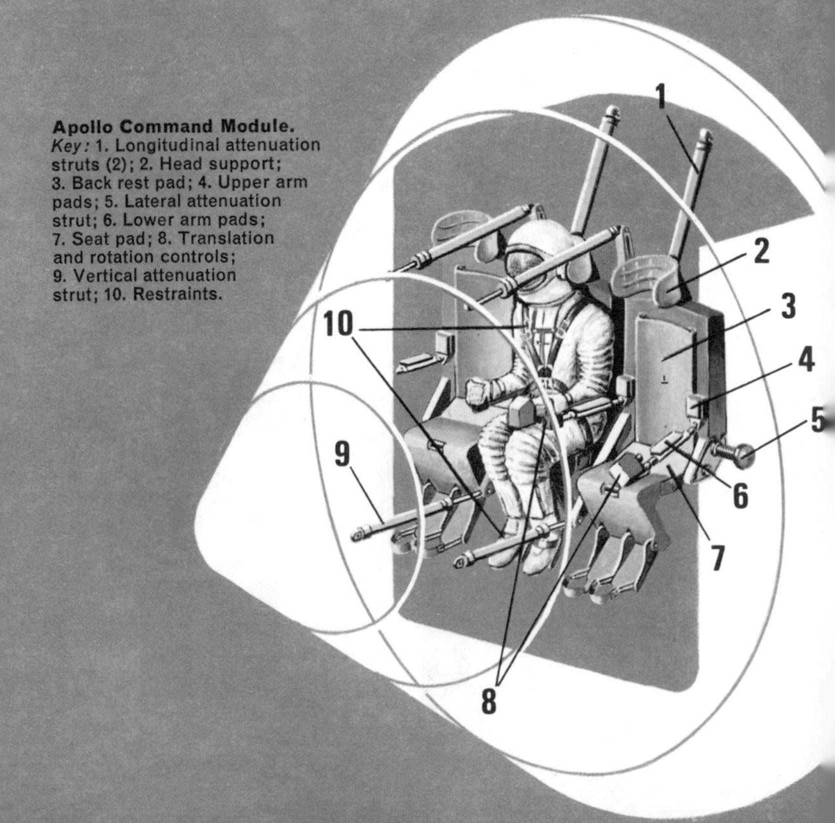

Apollo Command Module.
Key: 1. Longitudinal attenuation struts (2); 2. Head support; 3. Back rest pad; 4. Upper arm pads; 5. Lateral attenuation strut; 6. Lower arm pads; 7. Seat pad; 8. Translation and rotation controls; 9. Vertical attenuation strut; 10. Restraints.

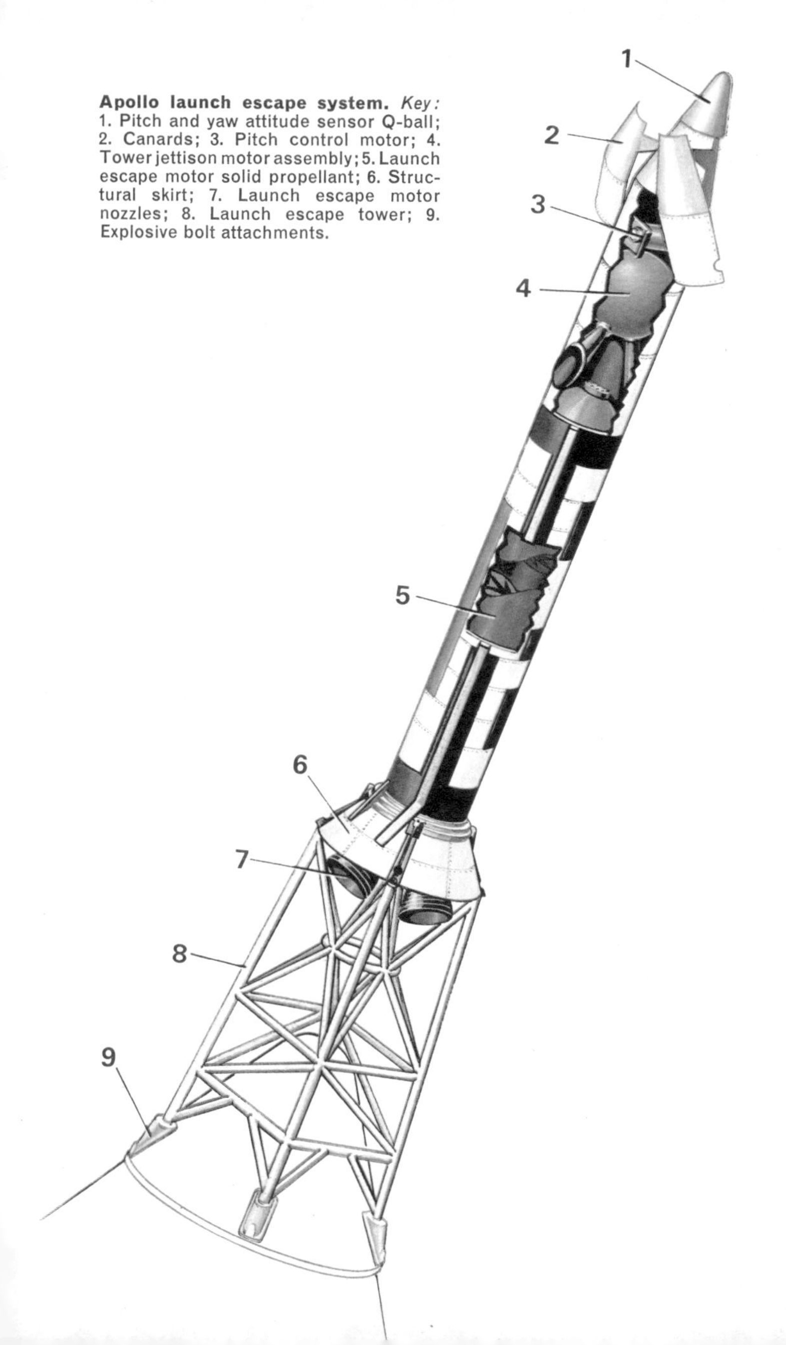

Apollo launch escape system. *Key:* 1. Pitch and yaw attitude sensor Q-ball; 2. Canards; 3. Pitch control motor; 4. Tower jettison motor assembly; 5. Launch escape motor solid propellant; 6. Structural skirt; 7. Launch escape motor nozzles; 8. Launch escape tower; 9. Explosive bolt attachments.

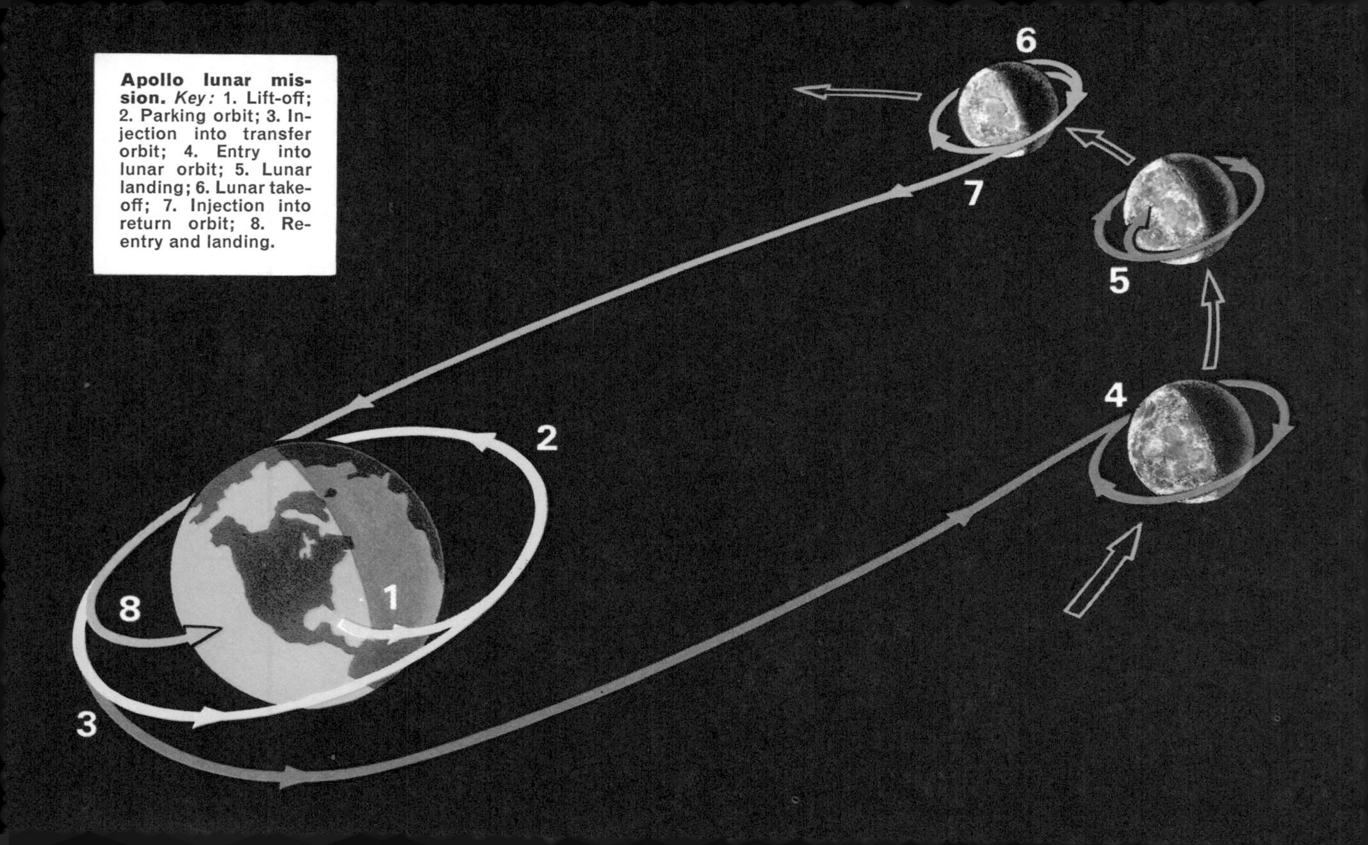

Apollo lunar mission. *Key:* 1. Lift-off; 2. Parking orbit; 3. Injection into transfer orbit; 4. Entry into lunar orbit; 5. Lunar landing; 6. Lunar take-off; 7. Injection into return orbit; 8. Re-entry and landing.

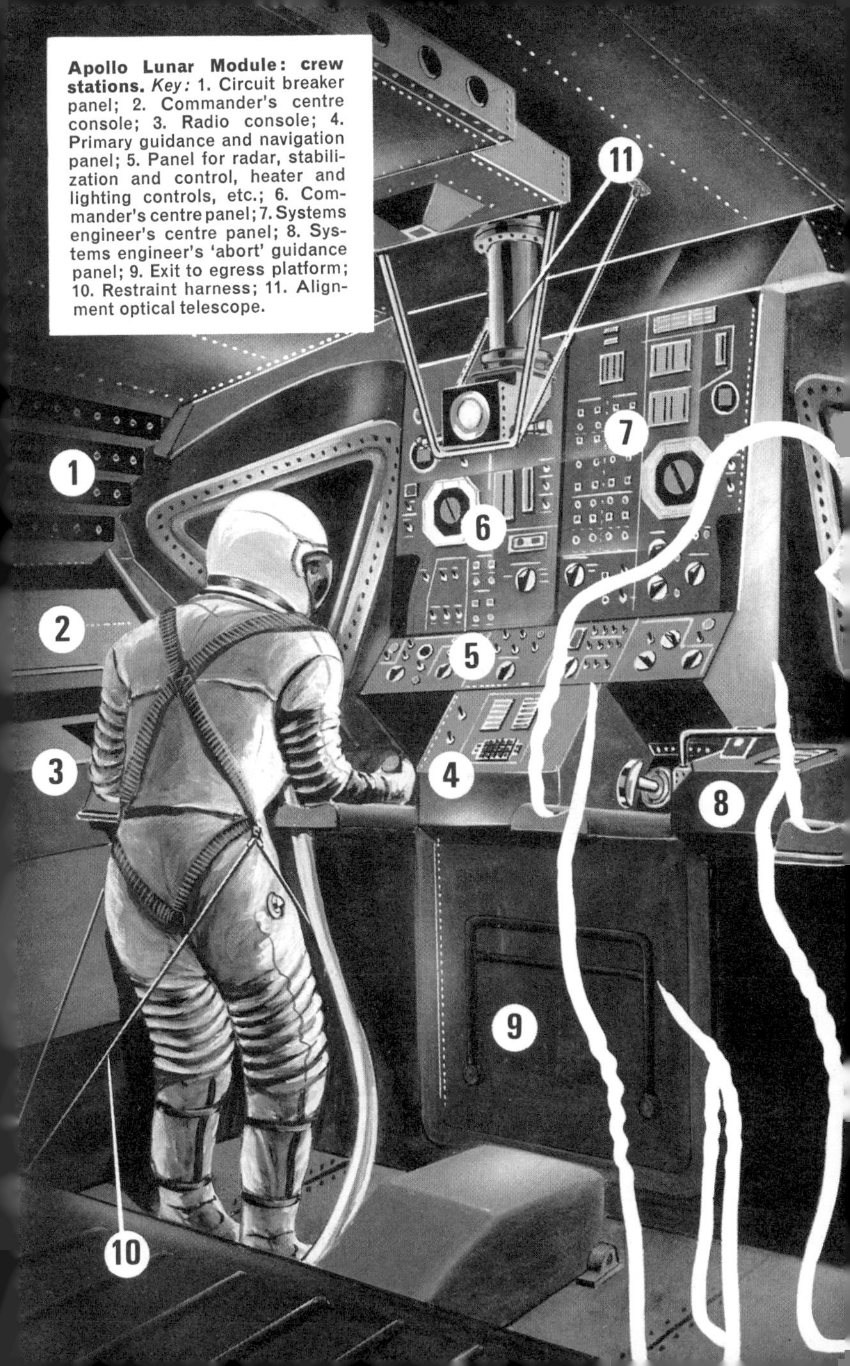

Apollo Lunar Module: crew stations. *Key:* 1. Circuit breaker panel; 2. Commander's centre console; 3. Radio console; 4. Primary guidance and navigation panel; 5. Panel for radar, stabilization and control, heater and lighting controls, etc.; 6. Commander's centre panel; 7. Systems engineer's centre panel; 8. Systems engineer's 'abort' guidance panel; 9. Exit to egress platform; 10. Restraint harness; 11. Alignment optical telescope.

(Below) Apollo lunar module descent stage (landing gear folded). *Key:* 1. Aft interstage fitting; 2. Fuel tank; 3. Engine mounting; 4. P.LSS, S-band antenna storage bay; 5. Structural skin; 6. Descent engine; 7. Insulation; 8. Thermal shield; 9. Forward interstage fitting; 10. Oxidizer tank; 11. Fuel tank; 12. Oxygen tank; 13. Helium tank/cryogenic; 14. Descent engine skirt; 15. Truss assembly (landing gear); 16. Secondary strut (landing gear); 17. Pad (landing gear); 18. Landing radar antenna; 19. Primary strut (landing gear); 20. Lock assembly (landing gear); 21. Scientific equipment bay; 22. Gimbal ring; 23. Adapter attachment point (landing gear); 24. Outrigger (landing gear); 25. Oxidizer tank; 26. Water tank.

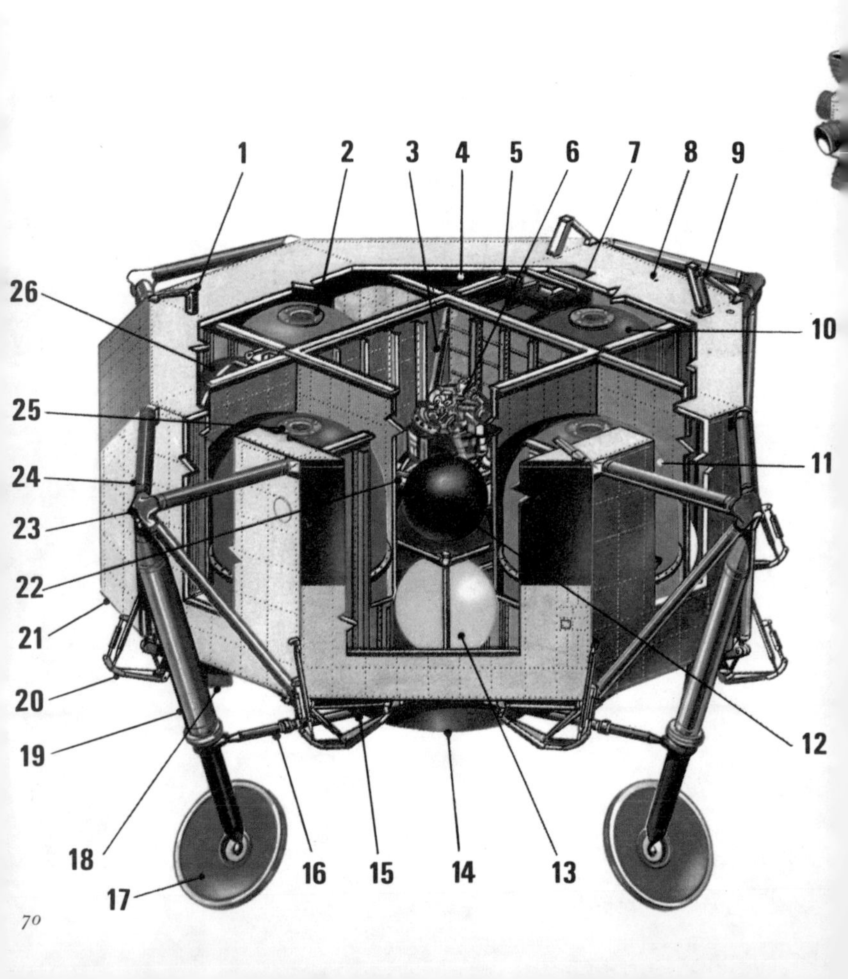

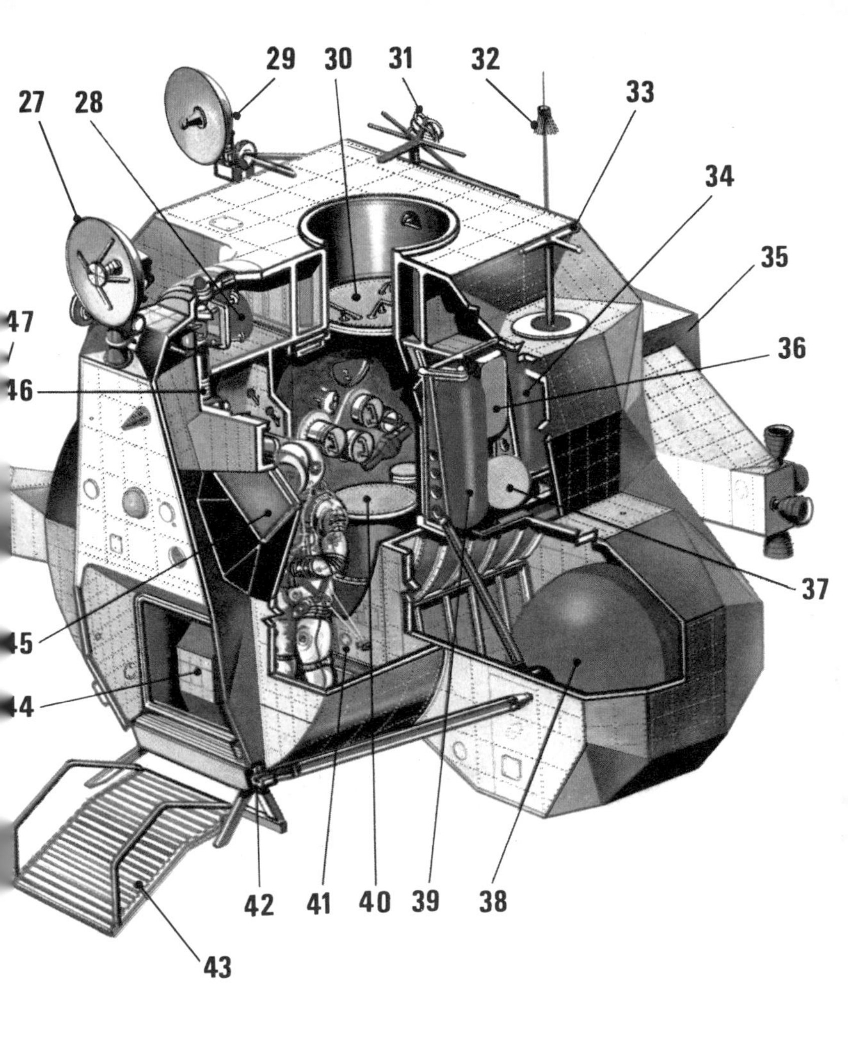

(Above) Apollo lunar module ascent stage. 27. Rendezvous radar antenna; 28. Inertial measuring unit; 29. S-band steerable antenna; 30. Docking hatch; 31. VHF antenna; 32. EVA antenna; 33. Docking target; 34. Fuel tank (RCS); 35. Aft equipment bay; 36. Helium pressure regulating module; 37. Helium tank; 38. Fuel tank; 39. Oxidizer tank (RCS); 40. Ascent engine cover; 41. Crew compartment; 42. Forward interstage fitting; 43. Egress platform; 44. Ingress/egress hatch; 45. Cabin window; 46. Alignment optical telescope; 47. RCS thruster assembly.

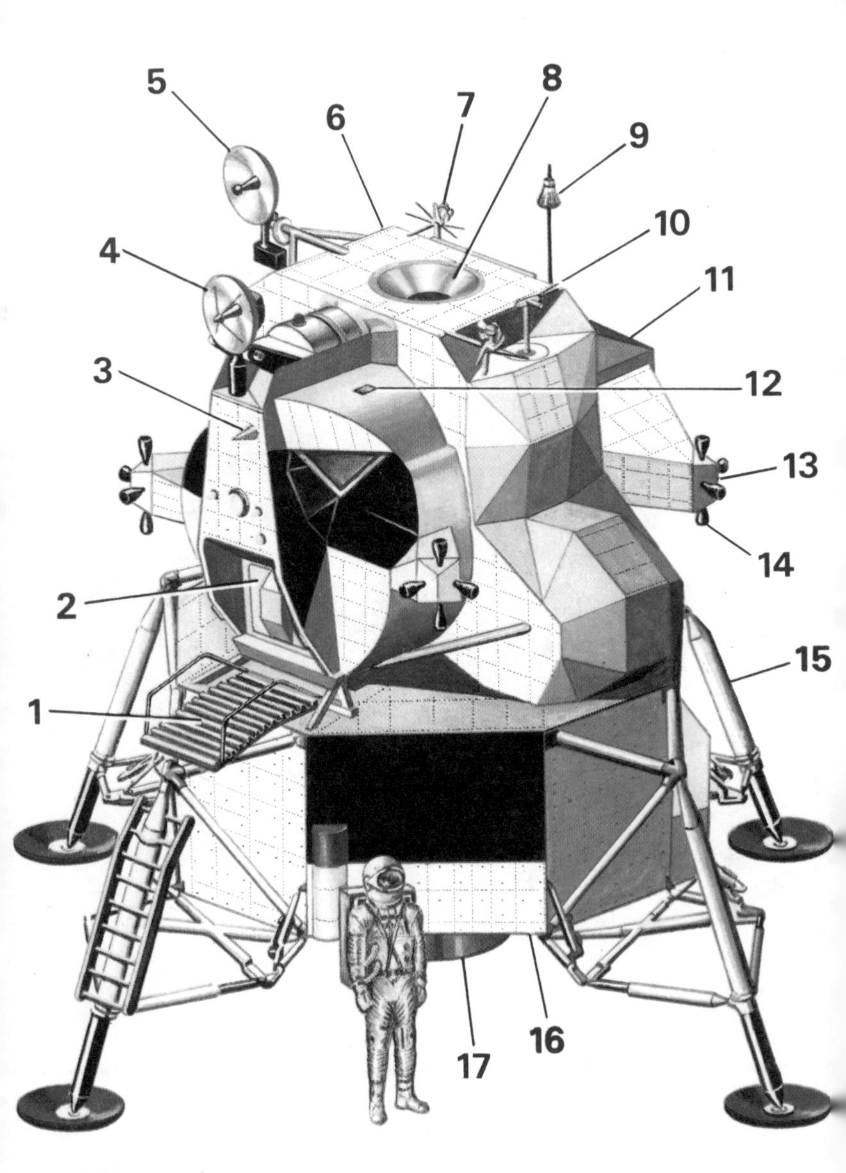

Apollo Lunar Module (complete). 1. Egress platform; 2. Ingress/egress hatch; 3. S-band inflight antenna (2); 4. Rendezvous radar antenna; 5. S-band steerable antenna; 6. Ascent stage; 7. VHF antenna (2); 8. Docking hatch; 9. EVA antenna; 10. Docking target; 11. Aft equipment bay; 12. Overhead docking window; 13. RCS thruster assembly; 14. RCS nozzle; 15. Landing gear; 16. Descent stage; 17. Descent engine skirt.

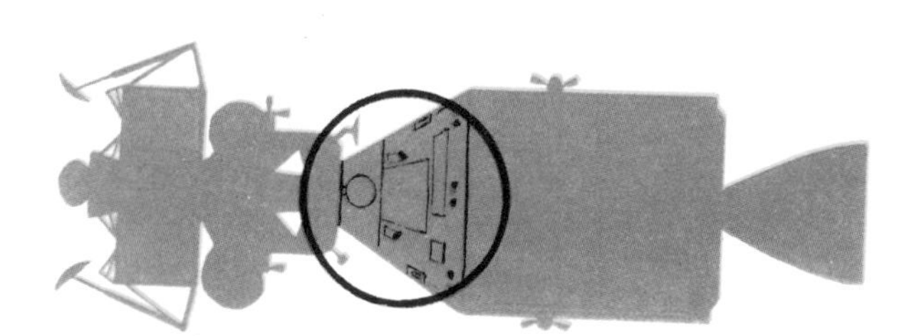

Apollo docking mechanism for linking command and service modules with lunar module following rendezvous manœuvres. *Key:* 1. Probe; 2. Probe extended; 3. Drogue; 4. Latch (12 positions).

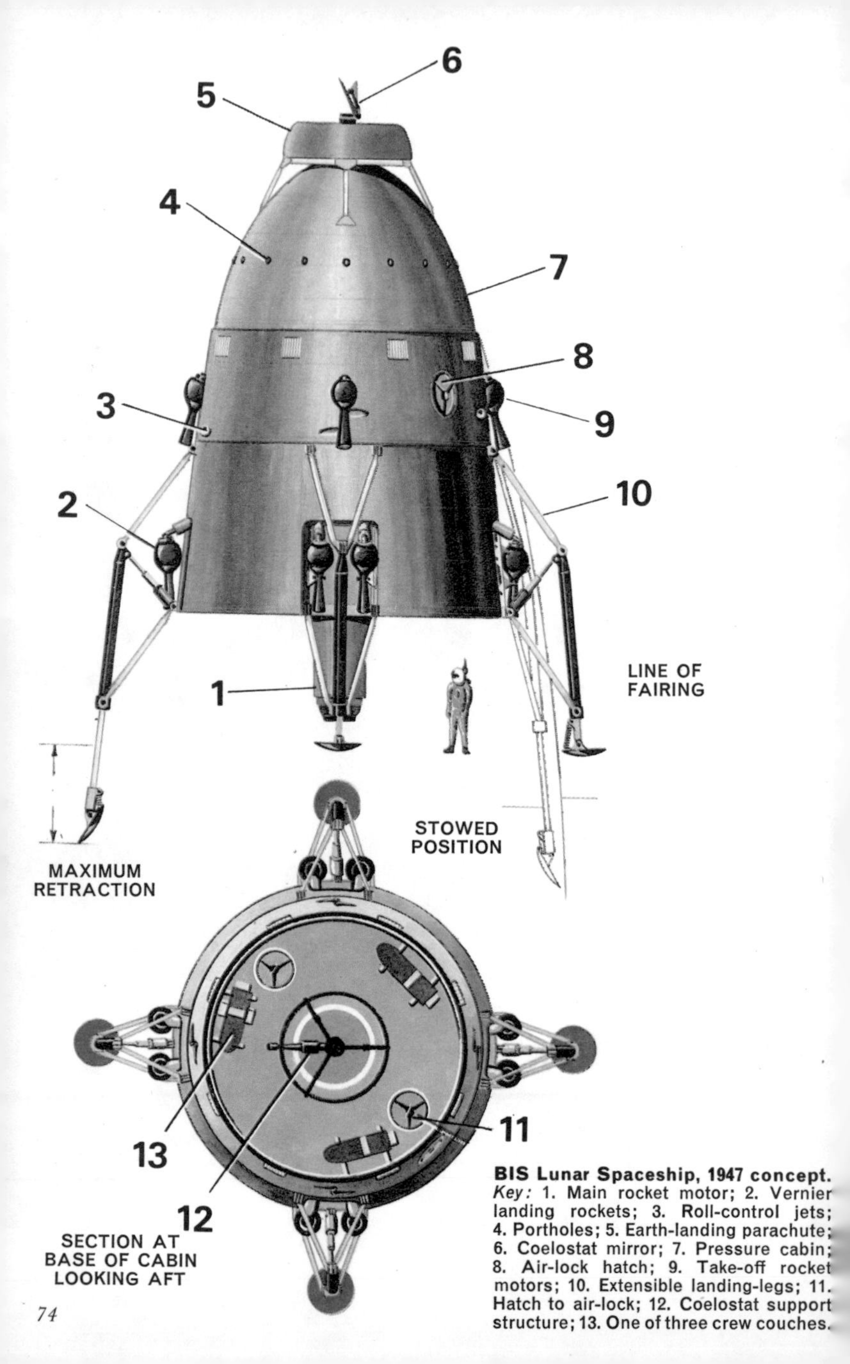

BIS Lunar Spaceship, 1947 concept. *Key:* 1. Main rocket motor; 2. Vernier landing rockets; 3. Roll-control jets; 4. Portholes; 5. Earth-landing parachute; 6. Coelostat mirror; 7. Pressure cabin; 8. Air-lock hatch; 9. Take-off rocket motors; 10. Extensible landing-legs; 11. Hatch to air-lock; 12. Coelostat support structure; 13. One of three crew couches.

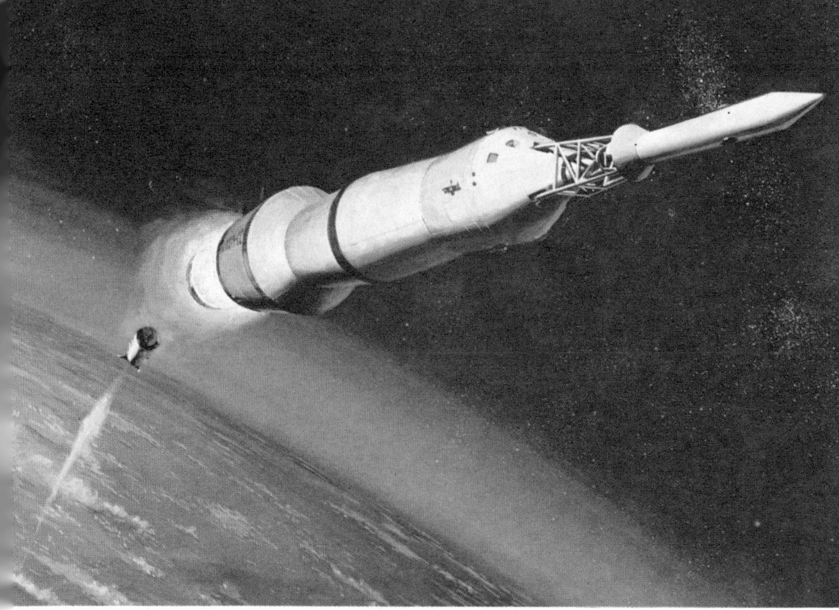

(Top) **Apollo stands ready for launching on the nose of Saturn 5; *(bottom)* first stage separates as stage two ignites 154 seconds after lift-off.**

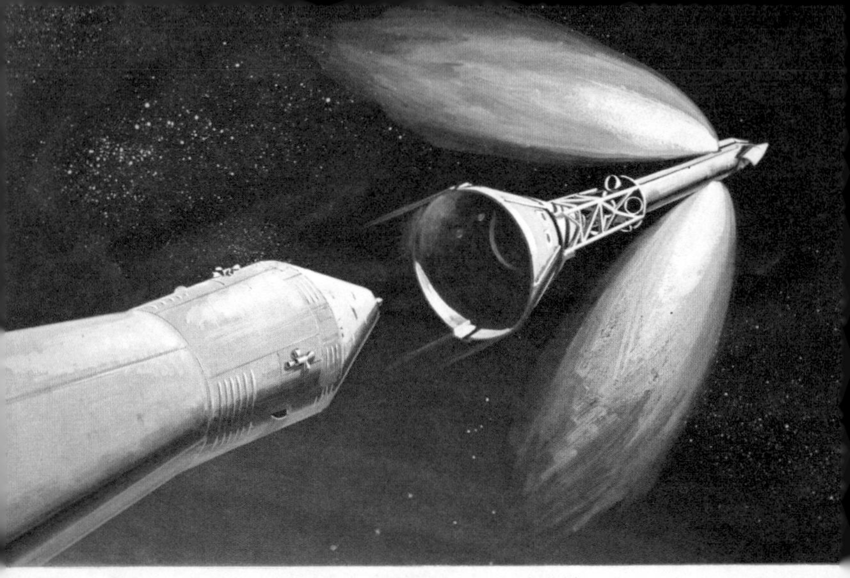

(Top) having cleared the atmosphere, escape tower and protective cone separate; *(bottom)* after delivering Apollo into a temporary parking orbit close to Earth, third stage fires again to inject vehicle into transfer orbit to the Moon.

(Top) command and service modules separate from third stage after adapter panels are blown free; *(bottom)* craft turns through 180° to line up with lunar module.

(Top) command module's docking mechanism engages docking collar of lunar module; *(bottom)* thrusters of service module pull craft free of carrier rocket's third stage.

(Top) parallel with moon's surface, service module's engine fires against flight direction to put combined vehicle into lunar orbit; *(bottom)* two astronauts transfer to lunar module ready to separate craft for landing.

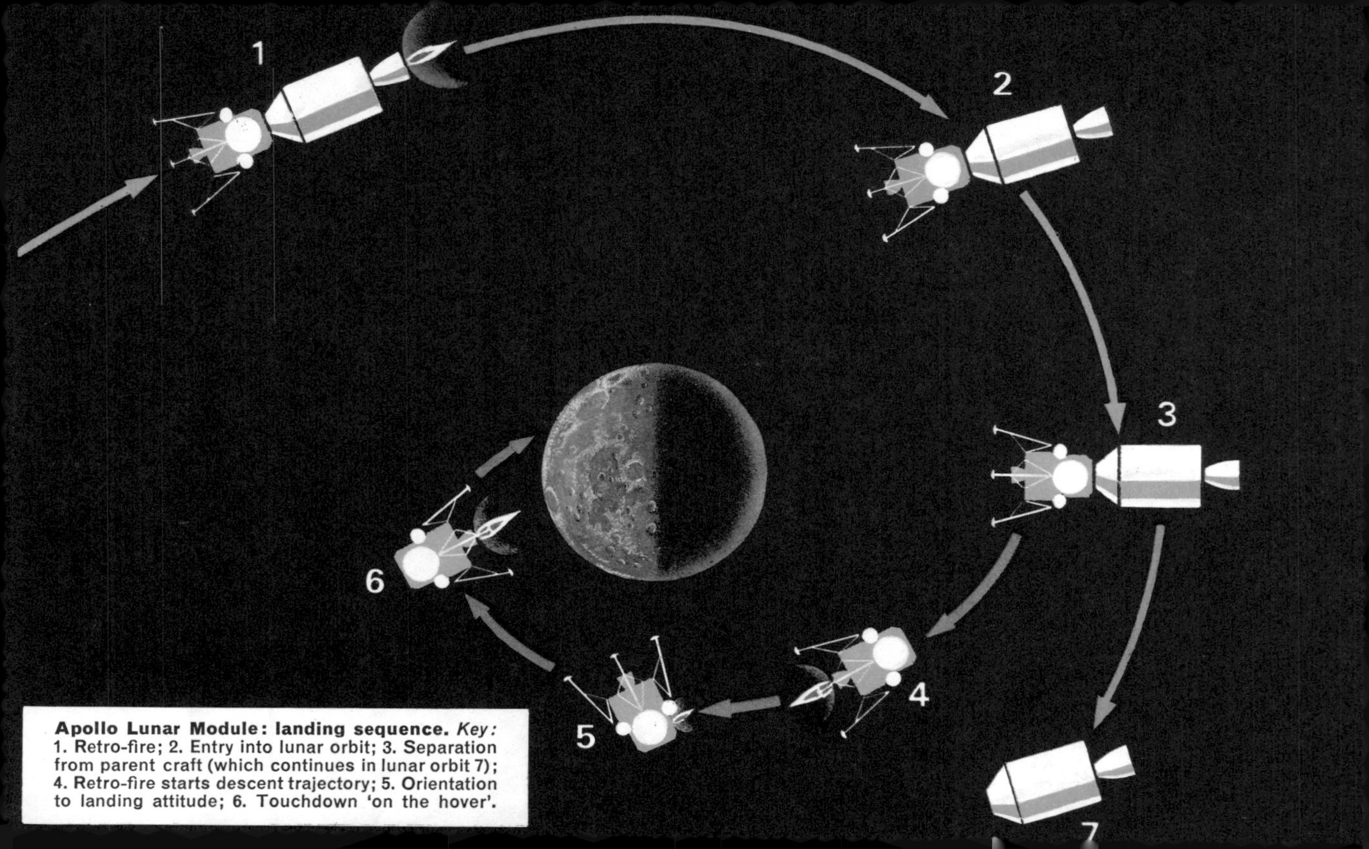

Apollo Lunar Module: landing sequence. *Key:* 1. Retro-fire; 2. Entry into lunar orbit; 3. Separation from parent craft (which continues in lunar orbit 7); 4. Retro-fire starts descent trajectory; 5. Orientation to landing attitude; 6. Touchdown 'on the hover'.

(Top) lunar module separates leaving one astronaut with parent craft in lunar orbit; *(bottom)* flare out to landing attitude with final rocket-supported hover to achieve smooth landing site.

(Top) leg-supported base section of lunar module becomes launch platform for ascent stage; *(bottom)* after rendezvous and docking with parent craft, return flight begins. Lunar module is left in lunar orbit.

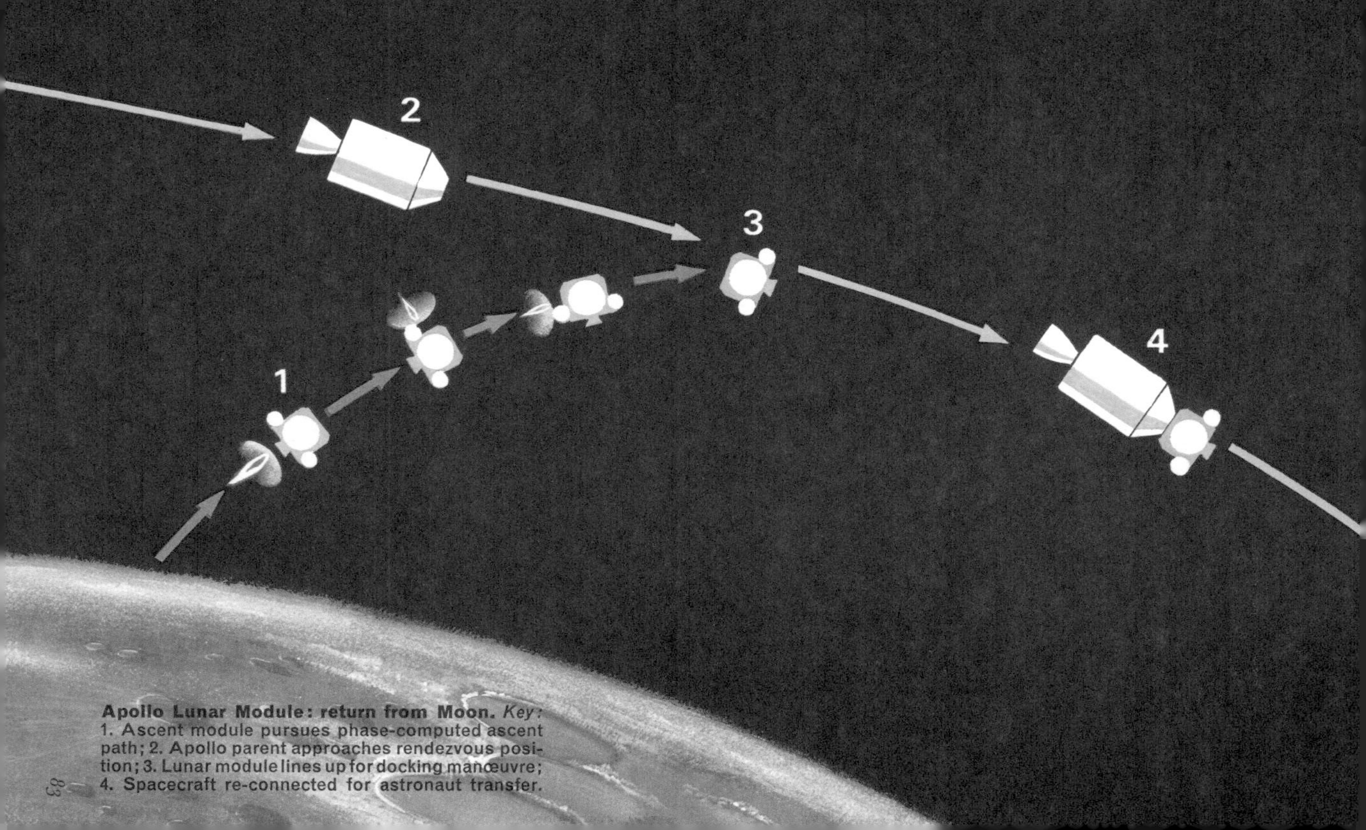

Apollo Lunar Module: return from Moon. *Key:* 1. Ascent module pursues phase-computed ascent path; 2. Apollo parent approaches rendezvous position; 3. Lunar module lines up for docking manœuvre; 4. Spacecraft re-connected for astronaut transfer.

Apollo re-entry sequence. *Key:* 1. Apollo on approach trajectory from Moon; 2. Turned by gas-jets; 3. Re-entry attitude with heat shield forward; 4. Re-entry heating.

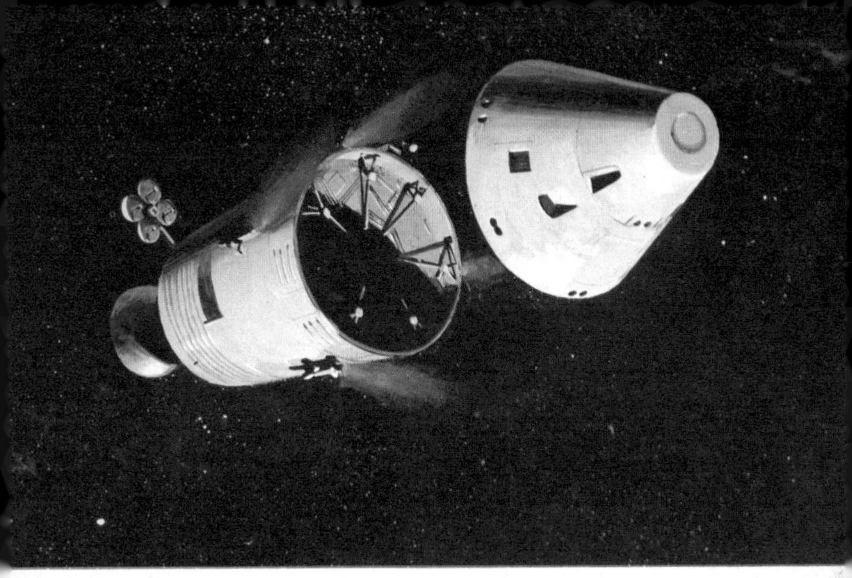

(Top) before re-entering Earth's atmosphere service module is jettisoned; *(bottom)* after command module is turned over ablative heat shield takes brunt of frictional heating.

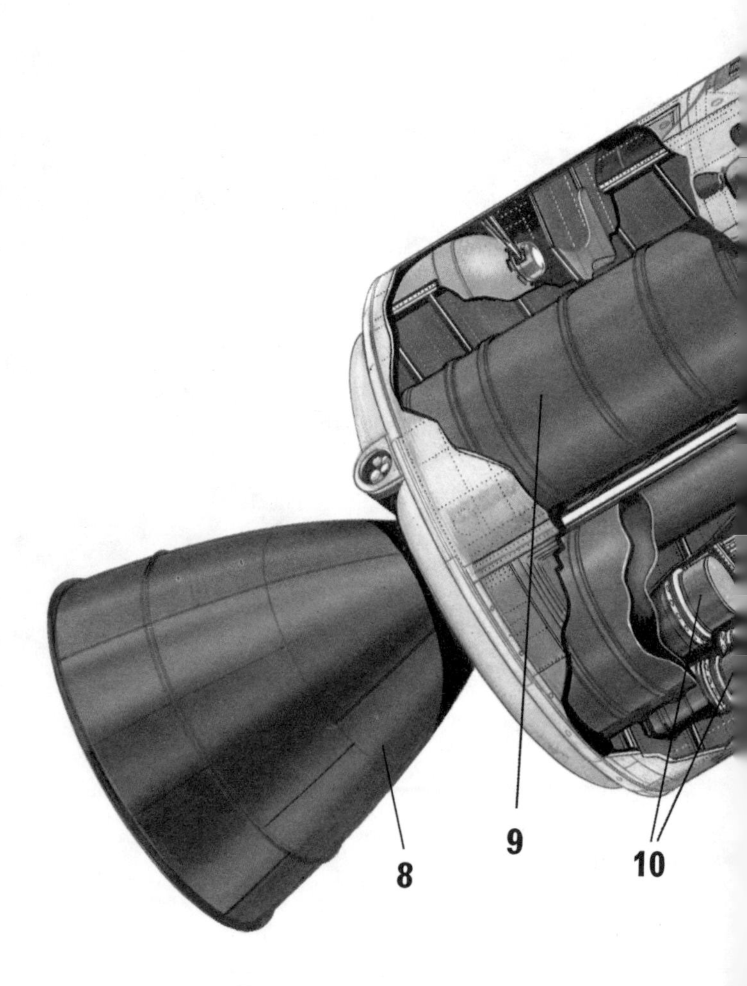
8
9
10

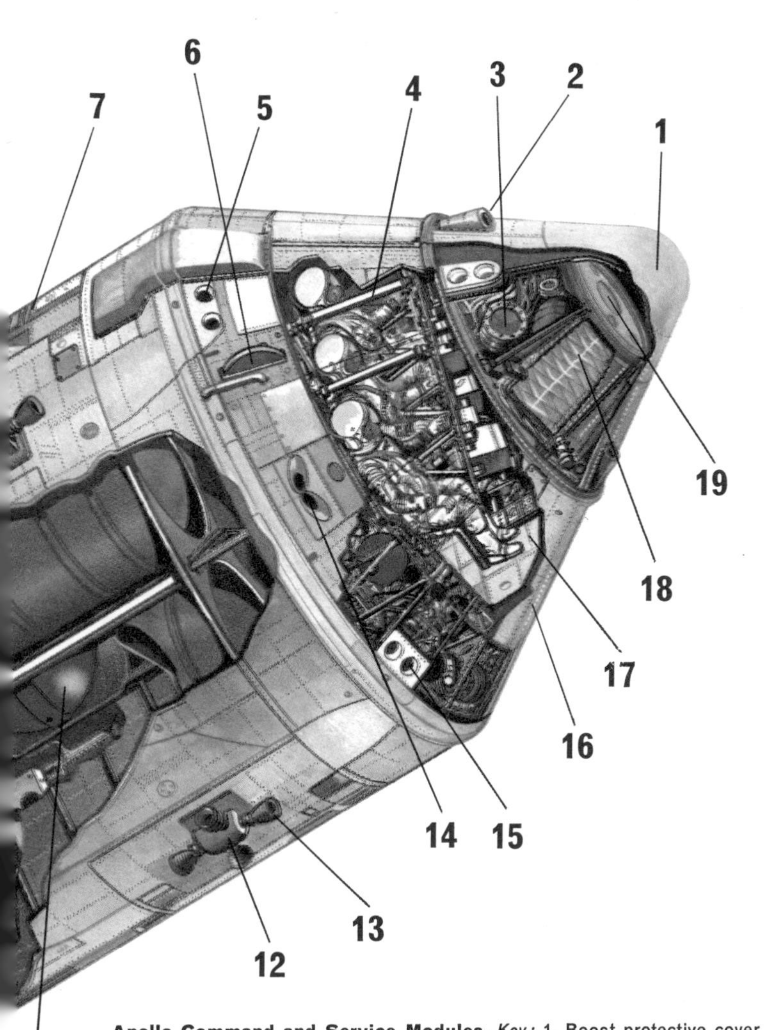

Apollo Command and Service Modules. *Key:* 1. Boost protective cover (apex section); 2. Explosive bolt attachments for launch escape tower; 3. Drogue parachutes and mortars; 4. Couch attenuation struts; 5. Pitch thrusters; 6. Antenna; 7. Environmental control system radiation core; 8. Service module propulsion engine nozzle; 9. Propellant tanks; 10. Fuel cells; 11. Helium tank; 12. Reaction control system quadrant; 13. Reaction control system engines; 14. Roll thrusters; 15. Yaw thrusters; 16. Aft boost cover; 17. Pressure cabin; 18. Parachute recovery system; and 19. Forward access tunnel.

Apollo recovery sequence. *Key:* 1. Parachute housing separates; 2. Drogue parachutes deploy; 3. Landing parachutes deploy; 4. Landing parachutes reefed; 5. Landing parachutes open; 6. Capsule awaits ocean retrieval.

(Top) command module's ringsail landing parachutes fully deployed; *(bottom)* rescue helicopters close in as Apollo astronauts splash-down.

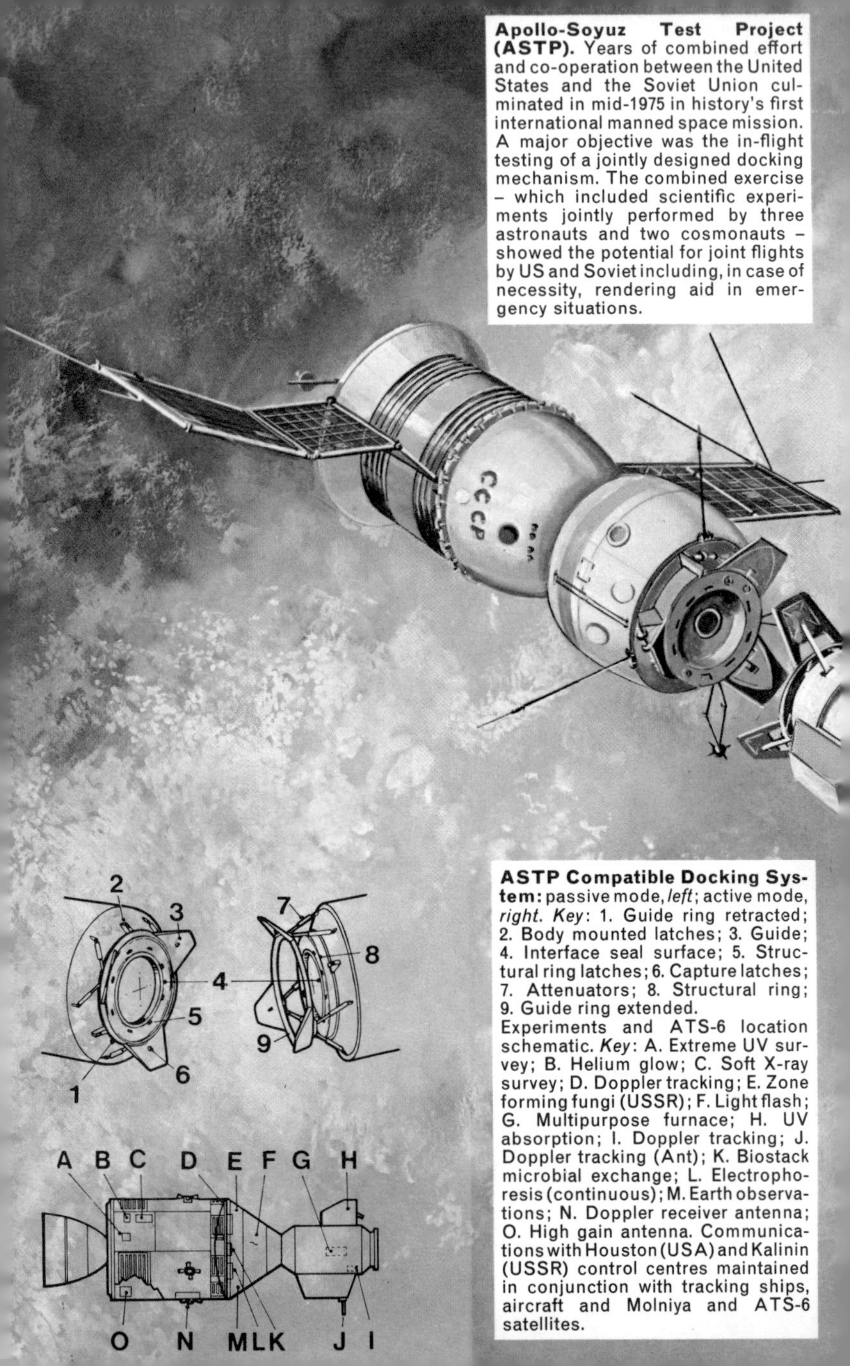

Apollo-Soyuz Test Project (ASTP). Years of combined effort and co-operation between the United States and the Soviet Union culminated in mid-1975 in history's first international manned space mission. A major objective was the in-flight testing of a jointly designed docking mechanism. The combined exercise – which included scientific experiments jointly performed by three astronauts and two cosmonauts – showed the potential for joint flights by US and Soviet including, in case of necessity, rendering aid in emergency situations.

ASTP Compatible Docking System: passive mode, *left*; active mode, *right*. *Key*: 1. Guide ring retracted; 2. Body mounted latches; 3. Guide; 4. Interface seal surface; 5. Structural ring latches; 6. Capture latches; 7. Attenuators; 8. Structural ring; 9. Guide ring extended.
Experiments and ATS-6 location schematic. *Key*: A. Extreme UV survey; B. Helium glow; C. Soft X-ray survey; D. Doppler tracking; E. Zone forming fungi (USSR); F. Light flash; G. Multipurpose furnace; H. UV absorption; I. Doppler tracking; J. Doppler tracking (Ant); K. Biostack microbial exchange; L. Electrophoresis (continuous); M. Earth observations; N. Doppler receiver antenna; O. High gain antenna. Communications with Houston (USA) and Kalinin (USSR) control centres maintained in conjunction with tracking ships, aircraft and Molniya and ATS-6 satellites.

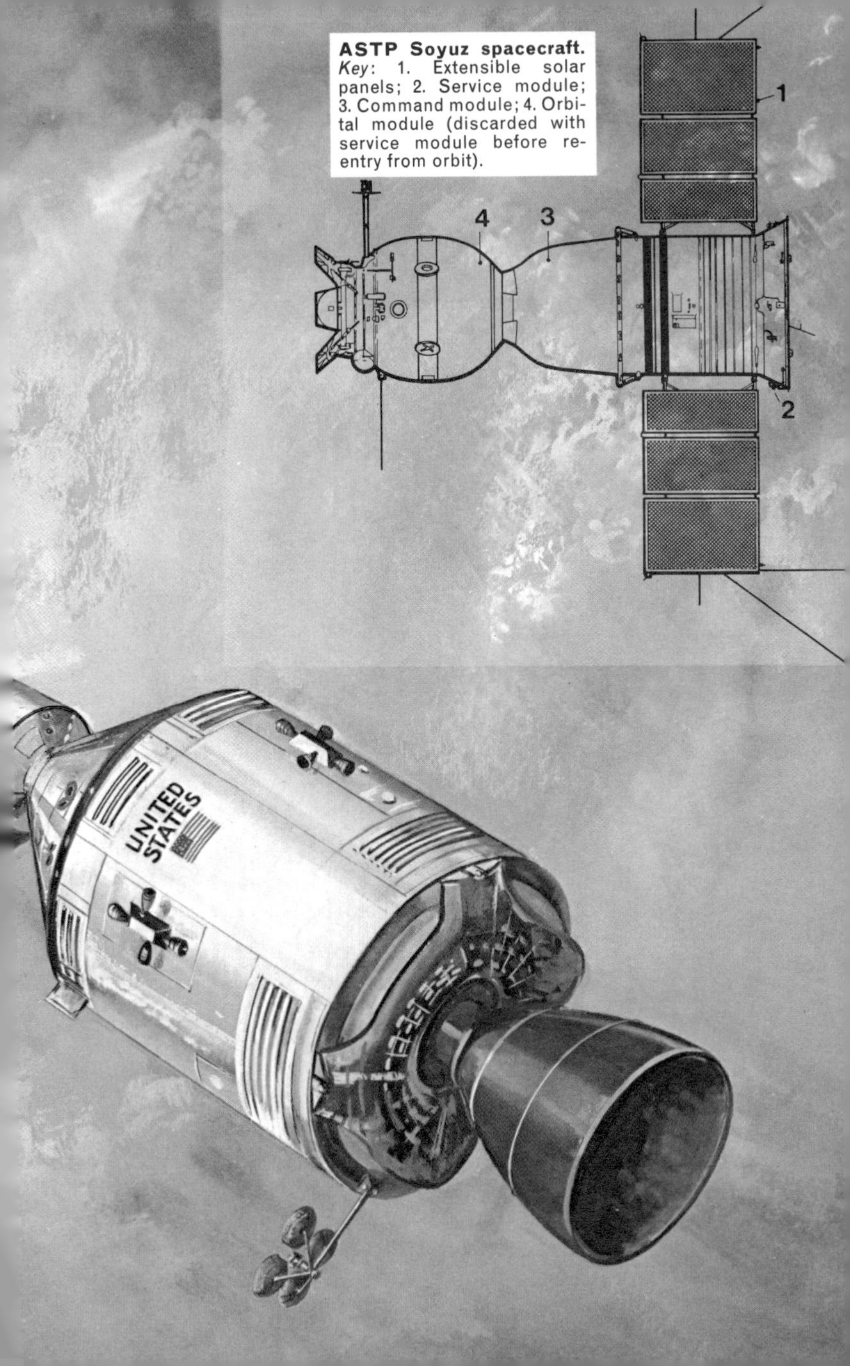

ASTP Soyuz spacecraft. *Key*: 1. Extensible solar panels; 2. Service module; 3. Command module; 4. Orbital module (discarded with service module before re-entry from orbit).

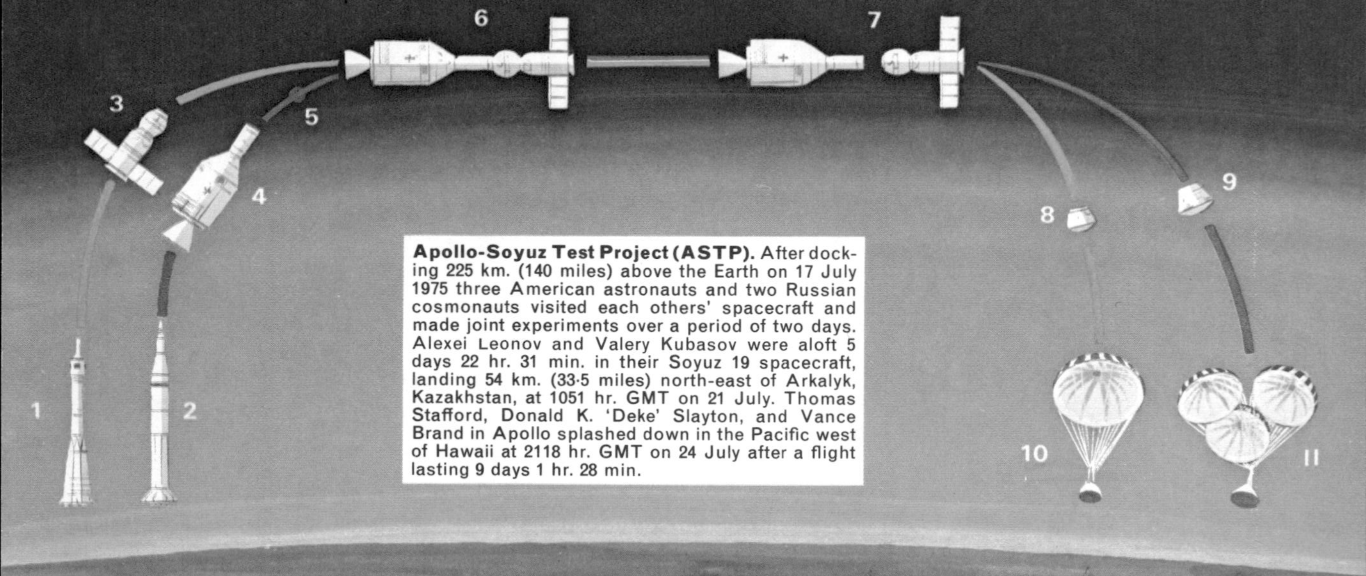

Apollo-Soyuz Test Project (ASTP). After docking 225 km. (140 miles) above the Earth on 17 July 1975 three American astronauts and two Russian cosmonauts visited each others' spacecraft and made joint experiments over a period of two days. Alexei Leonov and Valery Kubasov were aloft 5 days 22 hr. 31 min. in their Soyuz 19 spacecraft, landing 54 km. (33·5 miles) north-east of Arkalyk, Kazakhstan, at 1051 hr. GMT on 21 July. Thomas Stafford, Donald K. 'Deke' Slayton, and Vance Brand in Apollo splashed down in the Pacific west of Hawaii at 2118 hr. GMT on 24 July after a flight lasting 9 days 1 hr. 28 min.

Soviet Salyut 1 Space Station (1971). Soyuz three-man ferry approaches to dock with the station in this artist's impression based on Soviet official information.

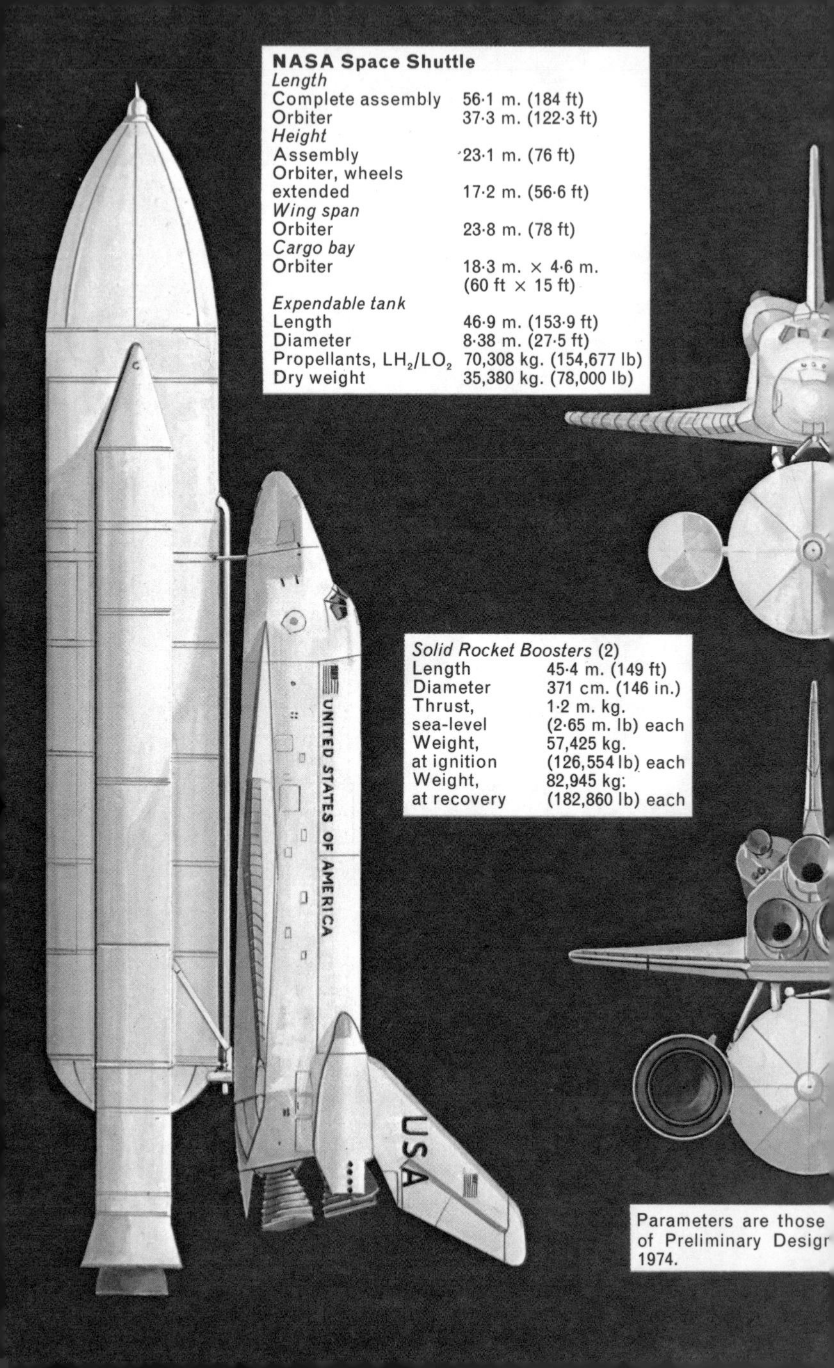
NASA Space Shuttle
Length
Complete assembly 56·1 m. (184 ft)
Orbiter 37·3 m. (122·3 ft)
Height
Assembly 23·1 m. (76 ft)
Orbiter, wheels extended 17·2 m. (56·6 ft)
Wing span
Orbiter 23·8 m. (78 ft)
Cargo bay
Orbiter 18·3 m. × 4·6 m. (60 ft × 15 ft)
Expendable tank
Length 46·9 m. (153·9 ft)
Diameter 8·38 m. (27·5 ft)
Propellants, LH_2/LO_2 70,308 kg. (154,677 lb)
Dry weight 35,380 kg. (78,000 lb)
UNITED STATES OF AMERICA
USA
Solid Rocket Boosters (2)
Length 45·4 m. (149 ft)
Diameter 371 cm. (146 in.)
Thrust, sea-level 1·2 m. kg. (2·65 m. lb) each
Weight, at ignition 57,425 kg. (126,554 lb) each
Weight, at recovery 82,945 kg. (182,860 lb) each
Parameters are those of Preliminary Design 1974.

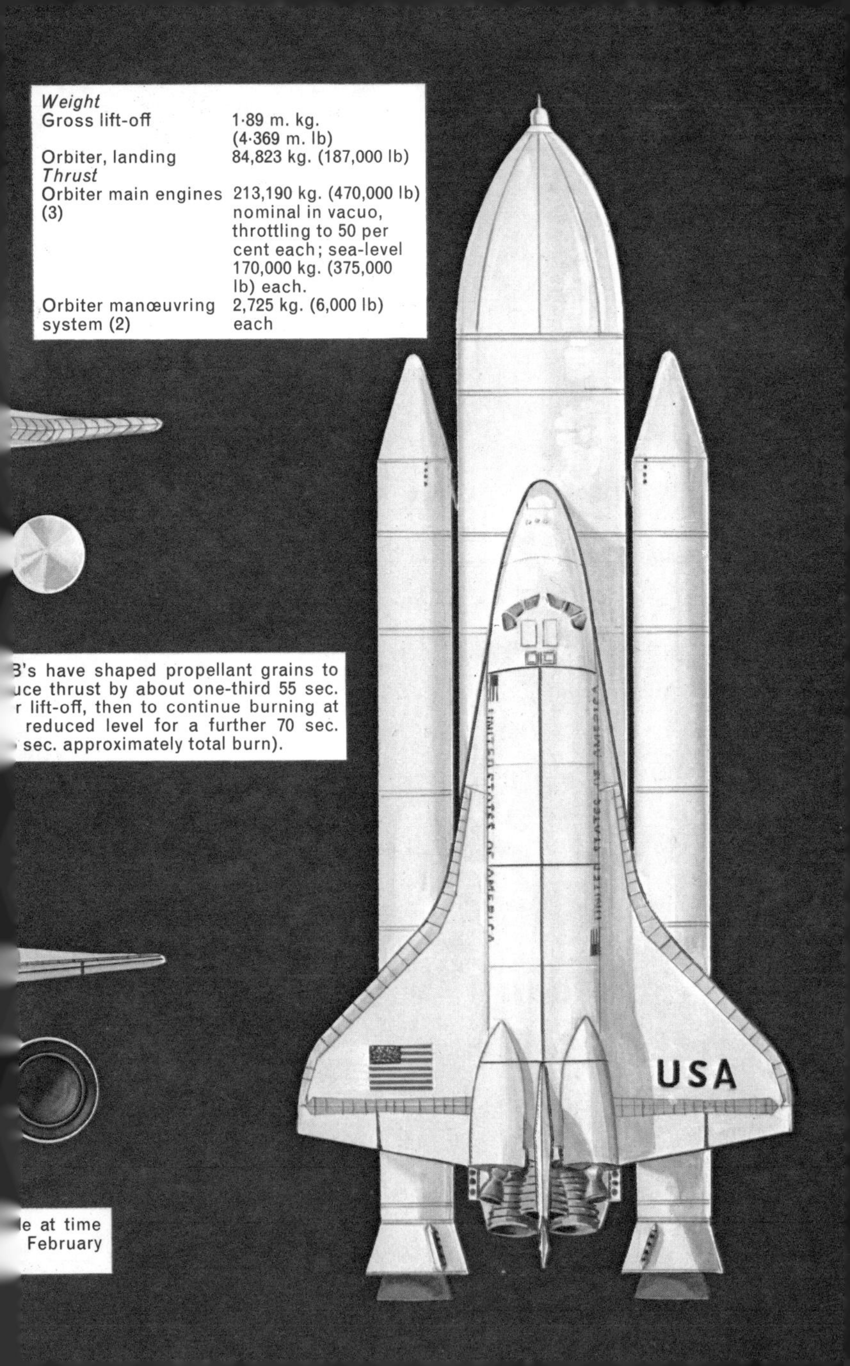

Weight
Gross lift-off 1·89 m. kg. (4·369 m. lb)
Orbiter, landing 84,823 kg. (187,000 lb)
Thrust
Orbiter main engines (3) 213,190 kg. (470,000 lb) nominal in vacuo, throttling to 50 per cent each; sea-level 170,000 kg. (375,000 lb) each.
Orbiter manœuvring system (2) 2,725 kg. (6,000 lb) each
B's have shaped propellant grains to uce thrust by about one-third 55 sec. r lift-off, then to continue burning at reduced level for a further 70 sec. sec. approximately total burn).
e at time February
UNITED STATES OF AMERICA
UNITED STATES OF AMERICA
USA

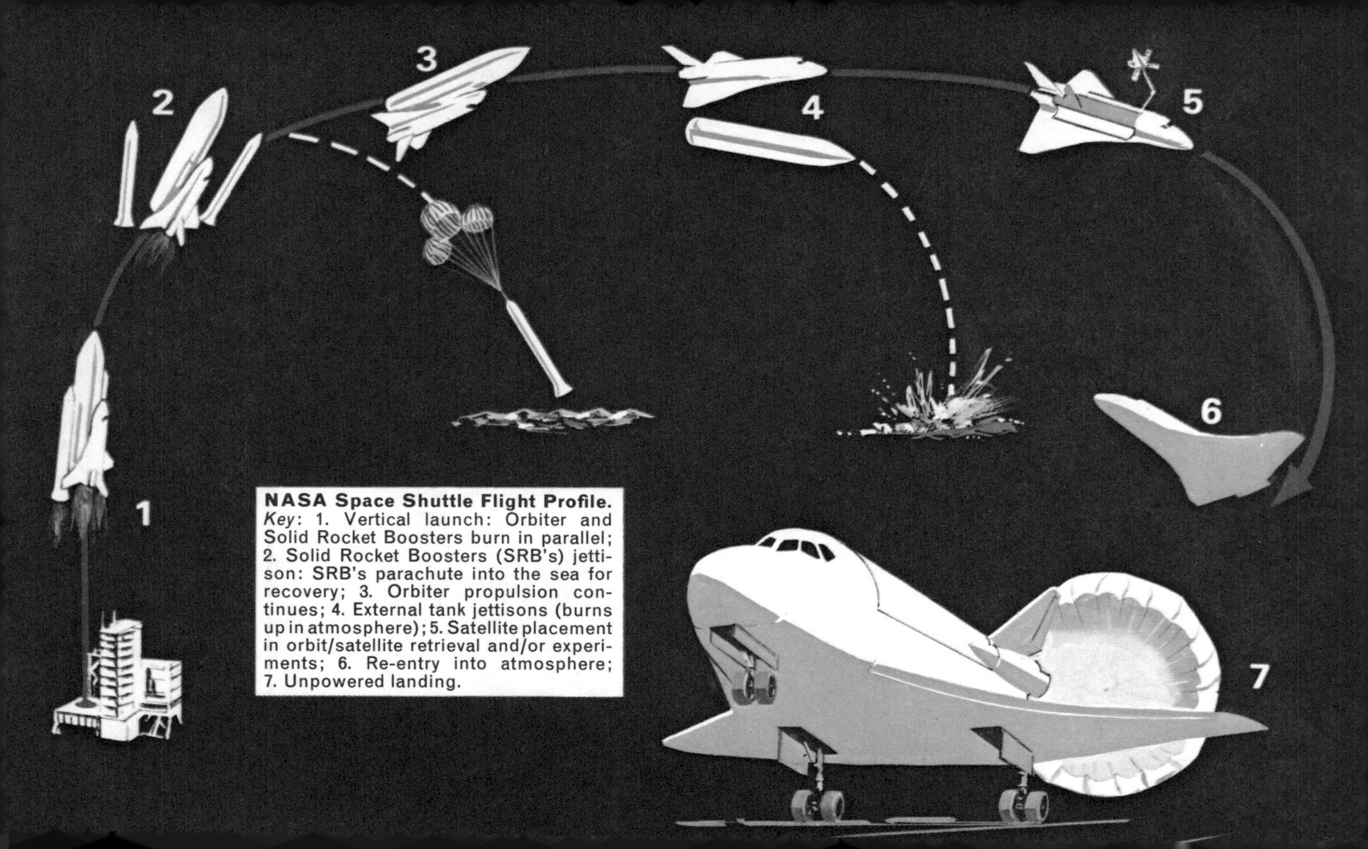

NASA Space Shuttle Flight Profile. *Key*: 1. Vertical launch: Orbiter and Solid Rocket Boosters burn in parallel; 2. Solid Rocket Boosters (SRB's) jettison: SRB's parachute into the sea for recovery; 3. Orbiter propulsion continues; 4. External tank jettisons (burns up in atmosphere); 5. Satellite placement in orbit/satellite retrieval and/or experiments; 6. Re-entry into atmosphere; 7. Unpowered landing.

CHAPTER ONE

RED STAR IN SPACE

The space age was barely eight hours old as our Vanguard airliner climbed away from London Airport *en route* for Barcelona. The date was 4 October 1957, and somewhere above the Earth the world's first satellite was bleeping its radio message signifying Russia's mastery of the new technique of satellite launching. Soviet scientists working in extreme secrecy had triumphed ahead of America just forty-eight hours before the opening of an important international Space Congress, on which delegates were converging from all parts of the world.

At the time none of us knew where this object – a 23-in. sphere of polished aluminium with four long whip aerials – had been launched. However, in a triumphant communiqué, *Tass* the official Soviet news agency had announced its weight as 83·6 kg. (184 lb); incredibly, this was 163 lb heavier than the satellite which America was preparing for the forthcoming International Geophysical Year. Although a picture of the satellite, Sputnik 1, was soon to appear in the Soviet Press, there was no photograph of the actual launching or indeed any details of the multi-stage rocket responsible for its appearance in space.

As we flew on towards Spain one event of the previous few months became a talking-point among the British delegates and pressmen aboard that plane. Barely two months before, in a few guarded sentences, *Tass* had announced the launching of an intercontinental ballistic missile (ICBM). The communiqué merely stated that '. . . a super long-distance intercontinental multi-stage ballistic rocket flew at an . . . unprecedented altitude . . . and landed in the target area'. Again there were no details of the rocket, nor was it stated how far it had travelled. But clearly here was a basis for the launch vehicle of Sputnik 1 which now circled our planet every 96 min., passing over every major continent.

Next morning, after we had signed in at the conference – the

Eighth Congress of the International Astronautical Federation – at last came the opportunity to question Soviet delegates on their triumph.

I well recall Academician Leonid Sedov's answers to eager questions put to him by Western delegates and correspondents as we stood talking in a group. Was it possible that Sputnik 1 was really so massive? Had *Tass* perhaps misplaced a decimal point? After all, America's unlaunched Vanguard satellite was 20 in. in diameter and weighed only 21 lb. Sedov's reply could scarcely be taken seriously; he said he had been away from the Soviet Union for a couple of weeks and therefore did not know which satellite had been launched.

Little did we then realize that within a month another sputnik would be in orbit with a dog-passenger, the announced weight being 508·3 kg. (1,120·8 lb).

Throughout the whole week of the Congress in Barcelona nothing more could be gleaned from the Russians than was contained in the original *Tass* statement, and all questions concerning the launch vehicle were met with polite but firm refusals to comment. For ten years the Russians maintained their strict code of silence.

As the first Russian satellites moved round the Earth in the last weeks of 1957, so in the West equipment was quickly improvised to track them using radio, radar, and optical methods. British observers were particularly active. Signals picked up by the British Broadcasting Corporation's listening station at Tatsfield enabled the satellite's time of closest approach to be estimated from signal strength and Doppler shift (analogous to the changing pitch of a whistle from a passing train). This was quickly supplemented by the active tracking of satellites at Jodrell Bank. Analysis of orbital data was made by the Royal Aircraft Establishment, Farnborough, in conjunction with many sources including the Royal Radar Establishment, Malvern, and the Admiralty Signals and Radar Establishment, Portsdown.

Stations in many parts of the world quickly began to gather vital data quite unconnected with telemetry signals giving data from instruments installed in the satellites which only Soviet scientists could interpret. How a satellite's path was distorted (perturbed) as it circled the Earth gave a clue to the Earth's non-uniform shape or unequal mass distribution, affecting the local strength of the Earth's gravity. New information was obtained on the nature of the radio-reflecting layers of the Earth's ionosphere by observing the scattering effect on the satellite's radio signals. The first reliable information was also obtained on the density of the upper atmosphere by observing the rate at which air drag caused the satellite to spiral ever closer to the Earth until it was burnt up by air-friction at a height of some 60 miles.

There was much to excite the scientist in those early days of satellite observation; but for the average citizen perhaps most thrilling of all was the ability to spot satellites with the naked eye as they made their transits before dawn and after dusk. Travelling at over 17,000 m.p.h. hundreds of miles up, they reflected sunlight from beyond the horizon revealing themselves to the ground observer as slow-moving 'stars' in the night sky.

Most conspicuous were the carrier rockets of Sputnik 1 and Sputnik 2. As the dog-cabin of the latter remained attached to its orbiting rocket stage, this had the greater length. Photographic and radar observations of these objects indicated lengths of 65 to 75 ft.

The last orbits of Sputnik 2 were particularly spectacular. As I watched from Coombe Hill in Surrey, it appeared to flash like a far-off beacon. This indicated that the long cylindrical vehicle was slowly turning over and over in space reflecting sunlight unevenly. The dog it contained, named Laika, had long since lost consciousness and died as its air supply gave out. Biomedical data were received from the animal experiment for about a week after the launching when radio transmissions ceased; the dog was fed and watered from an automatic dispenser within its tiny pressure cabin but no provision was made for its recovery. Sputnik 2 remained in orbit for five and a half months. As air drag brought it closer and closer to the Earth with each successive revolution, suddenly on 14 April 1958, it was gone in a spear of flame as air-friction heated it to incandescence.

This first experiment with a living creature in orbit gave an early clue to Russia's future intentions in manned spaceflight. For America, it was a particular challenge, for the rocketry involved was obviously on a scale greater than anything then available to the Western World. This view was further reinforced in May 1958 by the launching of Sputnik 3, a cone-shaped geophysical laboratory weighing fully 1,327 kg. (2,926 lb).

Anxiety that a serious missile-gap was developing between Russia and America led to the United States broadening the scope of military surveillance of Soviet affairs.

First definitive reports of Soviet missile activity, in fact, had begun to filter through to the West by 1950 following the repatriation of certain German rocket specialists who had been rounded up at the end of the Second World War and taken to Moscow. Once in the Soviet Union they had been formed into a Collective and given facilities in a near-derelict factory in the Moscow suburb of Khimki. They were not involved in the actual design of Soviet missiles but their talents were used to help solve basic problems in such fields as propulsion, structures, and guidance. Their only contact with the Soviet rocket programme came during regular technical meetings with Soviet specialists, and the Germans could only guess at the course that Soviet rocketry was taking. Later, the propulsion group was moved to an island in Seliger Lake near the source of the Volga. It was here that an engine of 264,000 lb thrust was engineered using liquid oxygen and kerosene as propellants with a chamber pressure of 880 p.s.i.

That Stalin himself was responsible for giving the Soviet rocket programme top priority in the immediate post-war years is now part of history. The importance of developing long-range missiles had been emphasized as early as 1946 by Marshal Zhigarev, then head of the Soviet Air Force. Should there ever be an American–Soviet war, he said, the V-2 would be useless. 'What we really need are long-range rockets of great reliability capable of hitting targets on the American continent.' The trend of thought within the Kremlin at this period has been fully discussed by Dr G. A. Tokaty, formerly Chief Rocket Scientist of the Soviet Government in Germany from 1945 to 1947.* Tokaty defected to the West in

* See *Astronautics in the Sixties* by Kenneth W. Gatland, Iliffe Books Ltd.

1947. He recalls that on 14 March 1947, he was present at a meeting with G. M. Malenkov and aircraft and rocket designers. Malenkov made it clear that the existing programme of V-2 development which had been taken over from the Germans did not conform to the long-term aims of the country – 'Our strategic needs are predetermined by the fact that our potential enemy is to be found thousands of miles away. . . .'

A day later, at a meeting of the Politburo and the Council of Ministers, Stalin made the aim even clearer. 'Under Hitler, German scientists have developed many interesting ideas,' he said with great seriousness. 'This Sanger project [a scheme for a rocket-powered sub-orbital bomber] seems to represent one of them. Such a rocket could have changed the fate of the war. Do you realize the tremendous strategic importance of machines of this kind? . . . The problem of developing transatlantic rockets is of extreme importance to us.'

This was the turning-point in Soviet ambitions. At this meeting Stalin personally suggested, and the Council of Ministers readily agreed, that a special State Commission should be established with the responsibility for managing the development of long-range rockets. Its initial members were: Colonel-General I. A. Serov (First Deputy Minister of the NKVD and Chairman); Professor Colonel G. A. Tokaty-Tokaev (Chief Rocket Scientist and Deputy Chairman, from the Soviet Air Force); Professor M. V. Keldysh (from the Ministry of Armaments); Professor M. A. Kishkin (from the Ministry of Aircraft Production), and Major-General Vassily I. Stalin (Marshal Stalin's son).

Meanwhile, the first task of the German Collective had been to assist in the launching of reconstructed German V-2s from a site which had been hastily improvised on the steppes of Kazakhstan, some 125 miles east of Stalingrad (now Volgograd). The first of these 12½-ton rockets, fired on 30 October 1947, travelled a distance of 185 miles and landed in the target area. A second launching on 13 November was much less successful. A guidance fault caused the rocket to oscillate wildly when it reached 500 ft, the fins were torn off and it plunged to destruction within sight of the launch crew.

On the evidence of Dr Tokaty, much-improved rockets of Soviet manufacture but retaining certain features of the V-2 were being produced in quantity in 1949–50. They were 'built under Soviet administrators, by Soviet workers, from Soviet materials, on Soviet soil' and bore the name Pobeda (Victory). Whereas the original German V-2 had a range of 190 miles, Pobedas could reach up to 560 miles. Examples first appeared in the Red Square military parade of November 1957 and in the West they are known by the NATO code-name Shyster.

According to Dr Tokaty, the first Rocket Divisions of the Soviet Armed Forces, equipped with V-2s and Pobedas, were formed in 1950–1. Upper-air research with rockets of the V-2 type started in the autumn of 1947; from 1949 it was continued with Pobedas.

Systematic research into the effects of spaceflight on living organisms began in 1949. From 1949 until 1952 dogs were rocketed to heights of some 60 miles, chiefly to test the effects of weightlessness. They travelled inside a pressure cabin installed in the nose-cone of Pobeda rockets which separated in flight and was parachute-recovered.

As more and more Intelligence reports filtered through to the West of Soviet activity at this site, steps were taken by the United States to set up radar stations on the perimeter of Soviet territory to monitor rocket activity which had begun to increase in the early 1950s north of the Caspian. By 1955 the US Air Force had put into operation a large fixed-beam radar at Diyarbakir, a mountain village in Turkey, which looked out across the Black Sea to the region in the Soviet Union where these experiments were in progress. The area is now famous as Kapustin Yar (literally Cabbage Crag), the Soviet cosmodrome from which early Cosmos satellites were launched into orbits inclined at 49° to the Equator.

However, long before the first Cosmos vehicle left this base in March 1962, missile testing had gone on almost continuously since the first V-2 launchings. In conjunction with other US radar stations on the perimeter of the Soviet Union, a disturbing picture of Russia's growing capability with long-range rockets was being built up by the mid 1950s.

Not only was it possible to spot missiles under test as they appeared above the radar horizon, but radio tracking was possible by homing on to the telemetry signals which carried data from various instruments embodied in the missiles. The electronic surveillance network gave information on the rockets' direction, speed, and altitude and it was possible to predict their impact points with considerable accuracy. The data showed that by the mid 1950s the Soviet Union was firing medium-range ballistic missiles, some of which impacted in the Kyzyl Kum Desert to the south-east of Kapustin Yar.

Meanwhile, the bigger rocket that Stalin insisted must be capable of reaching America, a weapon designed to carry a thermonuclear warhead that was both crude and heavy, was fast being developed by Soviet engineers.

The Americans had been faced with similar problems and as early as 1951 Convair (now General Dynamics) designed a massive rocket for purposes of H-bomb delivery, having nearly the same capability as the first-generation Soviet ICBM. In fact, had a Pentagon missile investigation committee not recommended cancellation of the project in 1953, on the grounds that the vehicle exceeded engineering experience, America could have matched Soviet early gains in space achievement. The proposed 200-ton multi-stage rocket had a first-stage thrust of 875,000 lb obtainable from a cluster of seven engines.

However, in military terms, abandonment of the big rocket was a wise decision. Towards the end of 1953 US physicists reported to the President the feasibility of a dry 'lightweight' thermonuclear warhead and a breakthrough in missile size was at hand.

Even so potential carriers were still heavier than the Atlas ICBM that eventually emerged. Convair in 1954 actually built a mock-up of a 650,000-lb thrust rocket designed round a cluster of five engines; it stood 90 ft tall and was 12 ft in diameter. When the operational SM-65 Atlas-D finally appeared, there were two 150,000-lb thrust engines in the jettisonable boost section and a central sustainer of 65,000-lb thrust. Lift-off weight was 255,000 lb.

The race to develop the hydrogen bomb led to Russia exploding her prototype device on 12 August 1953. Radiation sampling indicated the bomb was 'dry' though undoubtedly still unwieldy for missile installation. Although the Americans had exploded a thermonuclear device in the Pacific on 1 November 1952, it was 'as big as a house' and meant only for experimental purposes. It was not until 1954 that the Americans succeeded in testing a 'dry' H-bomb.

By then it must be presumed that the Soviet Union was well on the way with the task of marrying the H-bomb to a multi-stage rocket capable of launching it over thousands of miles. Thus, the Russians acquired an ICBM, cumbersome by American standards, but which could lift far heavier payloads than Atlas. Its use as a space booster was soon to be demonstrated.

It was in the mid 1950s that reports of another large cosmodrome under construction east of the Aral Sea began to reach the ears of Western Intelligence. This information helped to spur the development of another surveillance device, the celebrated U-2 reconnaissance aircraft.

Developed in great secrecy by Lockheed Aircraft Corporation, the U-2 was first used for high-altitude reconnaissance over the Soviet Union as early as the summer of 1956. Ostensibly it was used to gather weather information and this became a convenient 'cover story' for any infringement of Soviet airspace.

It has been reported that a U-2 spotted the first Soviet ICBM on its launch pad at the Baikonur/Tyuratam complex in August 1957, but as all information obtained by the Central Intelligence Agency (CIA) is secret, even today it is impossible to confirm the whole truth behind these early penetrations of the Iron Curtain. Before missiles were launched on test from this new site at target areas in the Central Pacific, they were directed towards the Kamchatka Peninsula in the Soviet far east.

However, the cat was fully out of the bag on May Day 1960 when Lieutenant Francis Gary Powers made his famous reconnaissance flight over the Soviet Union. We now know that on that day, as world leaders were gathering for a Summit Conference, a large multi-stage rocket was being prepared at the Baikonur cosmodrome. The U-2 flew at great height across the Afghan frontier on a course that took it over Stalinabad (now

Dushambe), south of Tashkent, and on across Aralsk and Baikonur. High above the Soviet cosmodrome its cameras looked down on the launch pads, blockhouses, and administrative buildings of Russia's most secret installations.

On flew the U-2 to the Chelyatinsk area and the approaches to Sverdlovsk where allegedly it was brought down by a Soviet surface-to-air missile (NATO code-name Guideline).

Fourteen days later the world knew what the CIA had long suspected; Russia was beginning a series of space experiments designed to put a man into orbit. According to *Tass*, the vehicle, known as Spacecraft 1, circled the Earth every 91·1 min. at a height ranging between 229 and 188 miles. It weighed 10,008 lb and contained a 'dummy cosmonaut'.

In the circumstances it was fortunate that this first test-vehicle did not contain a human being for when the capsule was commanded to re-enter the Earth's atmosphere it must have been wrongly orientated; instead of the retro-rocket reducing its speed it drove it into a higher orbit. Several parts of this vehicle were subsequently tracked as they circled the Earth, and in fact it was more than five years before the cabin section and its cosmonaut dummy re-entered the atmosphere on 15 October 1965.

What seems to have been a sub-orbital test for this spacecraft occurred in January 1960 when Russia began a series of test-firings of multi-stage launch vehicles impacting in the Central Pacific. Development of a 'nose-cone' which apparently remained intact after travelling 7,760 miles and reaching a maximum height of some 700 miles and a maximum speed of 16,150 m.p.h., not only suggested that Russia had solved the re-entry problem for ballistic missiles but that substantial progress had been made towards producing a re-entry vehicle for manned spaceflight.

After the initial firing on 20 January, Professor Boris Konstantinov, Director of the Physical and Technical Institute of the USSR Academy of Sciences, reported: 'Firstly, the discrepancy between the target and the point where the nose-cone, simulating the last stage of the rocket, hit the water represents only 1/6,000th part of the total range. Secondly, the two-way radio communication was maintained throughout the trajectory, including transit through the dense layers of the atmosphere over the target area. Thirdly, the thermal shield of the nose-cone fully protected it from destruction while it was moving at immense speed in the lower layers of the atmosphere. And lastly, a heavy satellite can carry a device making possible a safe descent through the whole atmosphere to the point of landing.'

Since that time Russia has used target areas in the Central Pacific for many different rocket tests. These launchings have been followed closely by the US Air Force and Navy who use radar and other sensors to monitor the ballistic re-entry bodies as they re-enter from space and approach their Pacific targets. Sometimes the last stage of Soviet rockets ascend 800 miles into space with a test-vehicle that separates in flight. The Americans, operating mainly from Hawaii, Kwajalein Atoll, and Johnston Island find the tests particularly useful for their own radar practice for the 'interception' of missile targets, Kwajalein Atoll and adjacent islands being the test centre for US Nike-Zeus and Sprint anti-missile missiles.

American ships and aircraft work together in these assignments monitoring the activities of Soviet tracking ships on station on the perimeter of target areas. When the re-entry body actually strikes the sea, the impact is recorded by sonar-devices giving precise measurements. After one particularly extensive series of firings between 13 September and 28 October 1961 the Soviet commander of tracking ships complained: 'US ships and aircraft do not leave the Soviet tracking ships alone for a single moment. American planes daily "buzz" the ships, coming closely over the mastheads and US ships manœuvre in such a way that they interfere with our work. Nevertheless, we quietly get on with our business.'

In the newspaper *Red Star* the unidentified commander claimed that the rockets landed accurately in the predetermined area after travelling some 7,500 miles. 'We worked calmly in the target area,' he said. 'We made contact with the rockets precisely at the appointed time and carried out measurements without any fear that they might deviate from their courses.' Accuracy was confirmed 'by very accurate measuring instruments in our ships'.

Following the mishap to the first Soviet spacecraft and its dummy cosmonaut, four more launchings were to take place before it was considered safe to entrust a man to orbital flight. The first of these occurred on 19 August 1960, when a 10,140-lb craft began orbiting the Earth with the dogs Strelka and Belka as passengers. Also in the cabin were smaller animals, insects, and plants.

The dogs occupied a pressurized container attached to the chassis of the ejection seat. The flight lasted just over a day and was entirely successful. Correctly orientated at a backward inclined angle for retro-rocket firing on the seventeenth orbit, the re-entry capsule separated from the instrument module as planned; after penetrating the atmosphere the dog-container was successfully ejected on the seat chassis for parachute recovery. Thus, the mission was a complete rehearsal for the subsequent Vostok manned missions.

After the flight the Russians released their first guarded statement on the vehicle. It included the information that the re-entry capsule travelled nearly 7,000 miles from the moment of being ejected from orbit to the time it had dropped to about 23,000 ft. It was subjected to a maximum deceleration of some 10 g.

A hatch on the capsule was released by barometric relays at a height of 23,000 to 26,000 ft whereupon the dog-container was ejected. It landed at a speed of 20 to 26 ft/sec.

At the time it was explained that this was an emergency procedure used on this occasion to test the escape system before manned spaceflights were attempted. Later, when human cosmonauts were being sent into orbit, its use was described as optional. In fact, it became standard practice for Vostok cosmonauts to eject in this way rather than remain in the re-entry capsule.

In the case of Spacecraft 2, it was stressed that the animals would have been safely recovered had they remained inside the capsule which came down near by. The landing, stated to have been only 6¼ miles from the preselected target point, was exceptionally accurate. An error of only 3 ft/sec. in velocity would have deflected the capsule over 30 miles. An error of about 300 ft in estimating the spacecraft's height over the Earth would have meant a deflection of nearly 3 miles, and an error of 1 min. of an arc in attitude a deviation of 30 to 40 miles.

First details of the spacecraft which was to become Vostok were also given, though at the time care was taken not to reveal any hint of its configuration. On the exterior were control jets and the 'orientation system's reserve high-pressure gas containers'. There were also 'the transducers of the scientific apparatus, the radio aerials, and experimental solar batteries. A system of heat insulation prevented the ship from burning up during the descent through the atmosphere. In the walls of the cabin were heat-resistant portholes and quickly detachable pressurized hatches.'

Apart from command signals from ground stations, the vehicle 'was guided automatically by means of a high-precision orbit control system'. Chemical and solar batteries powered the spacecraft's instruments. Solar batteries 'were placed on two half-discs 1 metre in diameter'. They remained locked on the Sun regardless of the spacecraft's attitude or position.

After the retro-rocket had started the vehicle on its descent path, the instrument module was separated and allowed to burn up in the atmosphere. This contained telemetry equipment; part of the scientific payload for studying cosmic rays and shortwave radiation from the Sun; equipment associated with guidance in orbit; heat-regulation apparatus and the retro-rocket system. Jettisoning of the instrument module greatly reduced the vehicle's mass, thereby 'reducing the demands on braking and recovery devices'. Exactly what these devices were was not disclosed but, as it turned out, they were simply a heat shield and parachutes.

One must acknowledge, however, the care with which Soviet engineers and scientists approached the problem of manned spaceflight. Years after the Strelka and Belka mission, it was revealed that their craft carried pieces of human skin which had been volunteered by doctors to determine the effects of radiation in space. The donors were Dr Kapichnikov, Dr Rybakov, and Dr Novikov of the Moscow Institute of Experimental Biology. Small flaps of skin, cut from their shoulders and thighs, were placed in sterilized glass bottles with a nutritive solution and dispatched to the cosmodrome. Similar control samples of skin were kept in the laboratory.

After spending a day in orbit, the flaps of skin were returned to the laboratory on 22 August 1960, where they were implanted together with the control samples back on the donors. It was found that the skin samples which had been in space began to knit on the eighth day, later than the control specimens which had remained in the laboratory. Nevertheless, together with the other bio-medical information that was being accumulated, it was good confirmation that a man could endure a day in orbit without harmful results.

Yet within four months the Soviet space programme was to receive another setback. On 1 December 1960, two more dogs Pchelka and Mushka, were launched in the 10,221-lb Spacecraft 3. The orbit achieved was lower than on the previous occasion, but the main fault appeared to have been the attitude of the spacecraft at the time of retro-fire. Having spent a day in orbit like the previous vehicle, it was delivered into an incorrect re-entry trajectory and the two dogs plunged to a fiery death as their capsule burnt up like a meteor.

How this affected the course of the Soviet space programme was never revealed. But it is probably significant that the next experiment on 9 March 1961 was limited to a single orbit. Only one dog Chernushka travelled in the 10,360-lb spacecraft and the orbit achieved was still closer to the Earth. Yet the tempo was building up. Just sixteen days later another launch vehicle left the Baikonur cosmodrome to orbit a 10,350-lb spacecraft with the dog Zvesdochka. Successfully recovered after 1 orbit, this was the final qualification test for the craft which carried Yuri Gagarin.

As the historic day of the world's first human spaceflight drew nearer the rigid secrecy that had prevailed in the Soviet Union showed signs of breaking down. It began with a rumour from Moscow that a Russian cosmonaut had been launched in secret on Friday, 7 April. A BUP message to London at midnight on 11 April quoted unidentified sources as suggesting that the cosmonaut was still undergoing physical examination by Soviet scientists, but that he was suffering from post-flight effects of a nature more emotional than physical. BUP continued:

> It is reported that the astronaut was a professional test pilot and son of a prominent aircraft designer. Russian and foreign journalists are maintaining a round-the-clock vigil, waiting for an official announcement. However, official sources still maintained an absolute silence, neither denying nor confirming the reports. Moscow Radio went off the air just before midnight without mentioning a spaceflight.

The story was further embellished from the Soviet Union by a French correspondent who suggested that the mystery cosmonaut was none other than Lieutenant-Colonel Vladimir Ilyushin, son of the famous Soviet aircraft designer. In fact there was no truth in these rumours. Mr Pierre Salinger, then Press Secretary to the White House, said that the US Government had no information whatsoever of a Soviet space launching on the day concerned. As for the young Ilyushin, he was traced to a Chinese health resort in Hangchow where he had arrived some time before. Having received a bad leg injury in a road accident the previous June, he had left the Moscow Traumatology Hospital and remained bedridden at home until the end of January 1961. He was still walking with a stick.

The world did not have long to wait for the real story. On the morning of 12 April 1961, at 0759 hr. BST (0959 hr., local Russian time) came the big announcement:

> The world's first spaceship, Vostok [East], with a man on board was launched into orbit from the Soviet Union on 12 April 1961. The pilot space-navigator of the satellite-spaceship Vostok is a citizen of the USSR, Flight Major Yuri Gagarin.
>
> The launching of the multi-stage space rocket was successful and, after attaining the first escape velocity and the separation of the last stage of the carrier rocket, the spaceship went into free flight on a round-the-Earth orbit. According to

the preliminary data*, the period of revolution of the satellite-spacecraft round the Earth is 89·1 min. The minimum distance from the Earth at perigee is 175 km. (108·7 miles) and the maximum at apogee is 302 km. (187·6 miles), and the angle of inclination of the orbit plane to the Equator is 65°4′. The spacecraft with the navigator weighs 4,725 kg. (10,418·6 lb), excluding the weight of the final stage of the carrier rocket.

Lift-off occurred at about 0907 hr. local time (0707 hr. BST) from the Baikonur cosmodrome, and 1 complete orbit was accomplished – matching the performance of the two previous dog-flights. The flight lasted 108 min., re-entry being initiated at 1025 hr. local time when the craft was over Africa. The landing was made 30 min. later near the village of Smelovaka in the Ternov District, near Saratov. Gagarin remained in the re-entry capsule which came to earth by parachute. The total distance travelled was 25,400 miles.

Much was made of the fact that Gagarin was an ordinary citizen of the Soviet Union. Born on 9 March 1934, the son of a collective farmer in the Gzhatsk District of Smolensk, he entered a secondary school in 1941. His studies were interrupted by the German invasion.

After the Second World War the Gagarin family moved to Gzhatsk, where the young Yuri's studies resumed in secondary school. In 1951, he graduated with honours from a vocational school in the town of Lyubersy, near Moscow. He received a foundryman's certificate, at the same time completing a course at an evening school. Yuri then studied at an industrial technical school in Saratov, on the Volga, from which he graduated, also with honours, in 1955.

It was while attending the industrial school that the man who was to become the world's first cosmonaut took his first steps in aviation. He attended the Saratov Aero Club where he completed a course of training in 1955, to enter the Air Force School

* These figures were subsequently revised by the Soviet authorities, as follows: perigee 112·4 miles; apogee 203 miles; period 89·1 min., and inclination 64°57′.

at Orenburg. Graduating with a first-class certificate two years later, he became a pilot in the Soviet Air Force. Following his acceptance for training as a cosmonaut he joined the Communist Party of the USSR in 1960. Married, his wife Valentia is one year his junior. At the time of the flight their two daughters Yelena and Galya were two years and one month old respectively.

Gagarin's impressions of the flight were later given in an interview with *Tass*. He said the sunlit side of the Earth was plainly visible; one could easily distinguish the shores and continents, islands and rivers, large areas of water and folds in the land. Over the Soviet Union he clearly saw 'the big squares of collective farms, and it was possible to distinguish ploughed land from meadows'. Prior to going into orbit, he had never flown above 15,000 m. (49,213 ft). From orbit he was able to discern the Earth's spherical shape. He described the sight as 'quite unique and very beautiful'. One could appreciate

> the remarkable colourful change from the light surface of the Earth to the completely black sky in which one can see the stars. The dividing line is very thin, like a belt of film surrounding the Earth's sphere. It has a delicate blue colour. And this transition from the blue to the dark is very gradual and lovely. It is difficult to put into words.

When he emerged from the Earth's shadow, Gagarin said the horizon looked different. 'There was a bright orange strip along it, which again passed into a blue hue and once again into dense black. What struck me most remarkably was how near the Earth seemed, even from a height of 187 miles.'

He did not see the Moon. The Sun was 'tens of times brighter than on Earth', the stars were bright and distinct, with far more contrast than when seen from the Earth.

Of his reactions to weightlessness, Gagarin said he felt excellent. Everything was easier to perform. 'This was understandable. Legs and arms weigh nothing. Objects were swimming in the cabin, and I did not sit in the seat as before but was suspended in mid-air. During the state of weightlessness I ate and drank.'

Gagarin recalled that he worked normally, noting his reactions in a log-book. 'Handwriting did not change, though the hand was weightless. But it is necessary to hold the writing-block or it would float away from the hands.' He maintained radio contact on various channels.

He reported his conviction that weightlessness had no effect on the ability to work. Transition from weightlessness to normal gravitational forces happened smoothly. 'I ceased to be suspended over the seat but eased myself into it.'

During a visit to Paris I was able to question Gagarin on various aspects of his pioneer mission. He confirmed that he remained in the capsule instead of ejecting from it for separate parachute recovery. The spherical capsule, he said, was orientated into the correct re-entry attitude for drag-braking by having its mass off-centre; he felt no ill-effects either during orbital flight or when re-entering the Earth's atmosphere.

At the Twenty-sixth Salon International de l'Aéronautique et de l'Espace in 1965, I examined with Gagarin a full-size Vostok re-entry capsule which seemed complete in every detail. The forward face of the sphere was charred and scored and I inquired if this was, in fact, one of the capsules which had been into space. After some difficulty with the interpreter he admitted the exhibit was only a realistic representation which had been heated to simulate the appearance after re-entry.

As Gagarin had been given a new appointment in 1963 unconnected with the cosmonaut training centre, I asked if he now looked upon himself more as an administrator than a cosmonaut. Unhesitatingly, he said he expected to go into space again, not once but several times.

A young Russian engineer by the name of Sergei played a part in testing the Vostok cabin before Gagarin's first flight. During simulation tests at a Soviet space centre he was sealed in the capsule as technicians at control consoles produced conditions that might arise during an actual spaceflight. One of the enforced emergency situations Sergei faced was large variations of cabin temperature. After a test, he said his clothes 'were dripping wet' and he lost over 13 lb in weight.

Vostok comprised two main components. The *spherical capsule* complete with cosmonaut's ejection seat, life-support equipment, and parachute landing system. This was attached to a *combined equipment module and retro-rocket*, which also contained instruments used during orbital flight. The equipment is described in detail in Chapter Four, but essentially it comprised the following:

1 Instruments and equipment necessary for maintaining human functions, including an air-conditioning system, a pressure-control system, food and water, and means for removing the body's waste products;
2 Flight-control equipment and manual controls;
3 Landing system;
4 Radio equipment for communicating with Earth;
5 A system for automatically recording the work of the instruments, radio telemetry system, and various sensors;
6 A television system for observing the cosmonaut at ground stations;
7 Instruments for recording physiological reactions of the cosmonaut;
8 Retro-rocket;
9 Orientation system;
10 Flight-control system;
11 Radio system for measuring orbital parameters;
12 Temperature-control system, and
13 Electrical supplies.

A summary of the flight of Vostok 1 follows:

12 April 1961 *Moscow time (hr.)*	*Event*
0907	Multi-stage rocket lifts-off at Baikonur cosmodrome.
0951	Automatic orientation system switched on. After emerging from Earth's shadow, solar sensor locks on the Sun and orientates Vostok 1 on to it.

0952	Vostok 1 passes over Cape Horn, moving north-east. Gagarin radios he is well and equipment working normally.
1015	Automatic programming unit commands sequence for switching on retro-rocket; Vostok 1 approaching Africa.
1025	Retro-rocket fired to begin descent trajectory.
1035	Vostok 1 re-entry capsule – now separate from instrument/retro-rocket module – enters dense layers of Earth's atmosphere.
1055	Capsule lands in predetermined area.

With Gagarin's pioneer orbit successfully completed, confidence grew in the ability to keep a man in space for a day, as the Russians had done the previous August with the dogs Strelka and Belka. On 6 August 1961 at 0900 hr. (Moscow time) another Russian citizen – Major Gherman Titov – lifted off in Vostok 2. While the flight was in progress, *Tass* gave the aims as 'research into the effects upon the human organism of a prolonged orbital flight to study man's working capacity during a sustained state of weightlessness'. According to preliminary data,* the communiqué continued, 'the spacecraft has been put into an orbit close to the calculated one, with the following orbital parameters: perigee 178 km. (110 miles), apogee 257 km. (159 miles), inclination to Equator 64°56′, and initial period 88·6 min. The spacecraft weighs 4,731 kg. (10,432 lb), excluding the final stage of the carrier rocket.'

The communiqué continued: 'Two-way radio communication is being maintained with Major Titov. The cosmonaut is transmitting on frequencies of 15,765, 20,006, and 143·625 Mc/s. A Signal transmitter also on board operates on a frequency of 19·995 Mc/s.

As the flight progressed frequent communiqués were broadcast by Moscow Radio. The cosmonaut was observed by television

* These figures were subsequently revised by the Soviet authorities as follows: perigee 114 miles, apogee 152 miles, period 88·46 min., inclination 64°5′36″.

to be operating his instruments normally, while a constant stream of bio-medical information on his condition was recorded at ground stations. During the third orbit he consumed his first meal in space, a three-course lunch in paste form squeezed from tubes. After resting for an hour during the fourth orbit, he conducted physical exercises and further work-tasks according to the pre-arranged flight plan. For an hour Titov 'tested Vostok's manual controls', reporting on the good behaviour of the vehicle when orientated by this method.

On the sixth orbit Titov exchanged greetings with Yuri Gagarin at the Command Centre and again swung the craft round its centre of mass using manual controls.

From 1830 hr. (Moscow time) to 0200 hr. the following day, the cosmonaut was scheduled to sleep. He reported: 'Before turning in I first fixed my hands, which seemed to be suspended in the air, and fell into a light slumber.' At first he slept fitfully and then fell into a deeper sleep, 'without dreams'. In contrast to sleeping on Earth he felt no necessity to turn from side to side. So good was Titov's sleep that it lasted 35 min. longer than envisaged in the programme, which gave rise to some anxiety at the Soviet Command Centre, all radio transmissions having been suspended during this period.

With Titov back in contact with ground stations, the cosmonaut breakfasted and resumed his work apparently unperturbed by the effects of weightlessness, making observations of the Earth and performing various scientific and engineering tasks. It was some time after the mission that Soviet scientists disclosed that bio-medical problems were not entirely absent. 'Under weightlessness,' Professor Vladimir Yazdovsky reported, 'unpleasant sensations of a vestibular character were felt with increasing strength, especially when the cosmonaut turned his head sharply or was observing swiftly moving objects.' These sensations diminished after Titov's sleeping period but did not wholly disappear until he encountered decelerative forces during re-entry. Professor Yazdovsky emphasized, however, that some people might be more sensitive to weightlessness than others.

Unlike Gagarin, Titov ejected from the Vostok 2 capsule to

make a separate descent by parachute. He told a Moscow Press Conference: 'After the retro-rocket fired and the craft entered its descent trajectory, I felt very fit and decided to try the second landing system. At low altitude the seat was ejected and my further descent was made by parachute. The ship landed successfully near by.' Titov completed 17 orbits of the Earth in 25 hr. 18 min., making his return in the Saratov area some 450 miles south-east of Moscow. During the flight he had travelled 703,150 km. (436,656 miles). Aircraft and helicopters were in the air to watch the return from orbit and to pick up the cosmonaut after he had landed.

It was a year and five days before the next Soviet manned spacecraft stood ready on its launch pad at the Baikonur cosmodrome. Lift-off of Vostok 3 came at 1130 hr. (Moscow time) on 11 August 1962; but the first announcement gave no clue to the key experiment that was being planned. The objectives were given as further information on human reactions to spaceflight, study of the cosmonaut's working ability under weightlessness, scientific observations, and 'the further perfection of systems of cosmic ships, and methods of their communication, guidance, and landing'.

No hint was given of feverish activities then proceeding at the cosmodrome in a bid to launch a second spacecraft into proximity with the first on the following day.

Vostok 3 carried Major Andrian Nikolayev in a launching said to have gone with clockwork precision. The orbit had the following initial parameters: perigee 111·8 miles, apogee 145·3 miles, period 88·3 min., and inclination 64°59′. Western observers watched the flight continue into the second day and then came the announcement over Moscow Radio that Vostok 4 had been launched with cosmonaut Lieutenant-Colonel Pavel Popovich. According to *Tass* launching had been effected 'from the same launch complex' of the cosmodrome as Vostok 3. Moreover, launching had been made 'within 1 second of the scheduled time'. The timing was vital for the object was to produce a separation of only 5 km. (3·1 miles) between the two ships at the point where Vostok 4 was injected into orbit.

Post-flight analysis of orbital data showed that the closest distance achieved was actually about 6·5 km. (4 miles). However, as the craft had no provision for matching orbits by rocket thrust, they rapidly drifted apart. At the start of the thirty-third orbit the separation distance was 850 km. (528 miles) and at the beginning of the sixty-fourth, 2,850 km. (1,770 miles). The orbit of Vostok 4 was remarkably similar to that of its companion. Figures for the initial orbit were quoted as perigee 111·8 miles, apogee 157·7 miles, period 88·5 min., and inclination 65°.

A wide range of experiments was performed during the group flight. The Russians reported obtaining an enormous amount of observational data, photographs, log-book records, and telemetry recordings. Communications were maintained with the two spacecraft by ground stations, and between the two cosmonauts in orbit. The men were required to get in touch with each other every half-hour, calling in turn. For the first time television broadcasts from spacecraft were transmitted to many countries. Since the flights lasted 4 and 3 days respectively, much of the data received had to do with bio-medical effects.

The cosmonauts had to conduct regular psychological, physiological, and vestibular tests. These involved keeping a careful watch on pulse, respiration frequency, appetite, adaptability to noise, vibration, overstrain, and weightlessness; also the ability to work and sleep.

Conditions within both spacecraft were stated to have remained within the following parameters: cabin pressure 755–775 mm.; oxygen percentage 21–25; carbon dioxide percentage 0·5 max., and temperature 13–26° C, the higher values representing conditions before lift-off.

Bio-medical measurements telemetered to Earth were:

a Electrocardiograms of heart condition and circulatory system;
b Pneumograms of breathing;
c Electroencephalograms reflecting conditions of central nervous system, also permitting analysis of sleep and alertness, fatigue and excitation;

d Skin-galvanic reactions by ohmic resistance of the skin to assist study of central nervous system; and

e Electro-oculogram, recording movements of the eyes to give objective information concerning vestibular disorders; in conjunction with pulse and breathing frequency, etc.

In the case of (*e*) tiny silver electrodes, placed at the outer corners of the cosmonauts' eyes, recorded bio-currents of the muscles of the eye-ball. Eye movements to left or right were recorded respectively as positive and negative currents. Skin-galvanic reflexes (*d*) were registered by measuring skin resistance by electrodes placed on the front and lower third of the cosmonaut's right shin. Changes occurred in the electrical potential of the skin and skin resistance as a result of various vegetative and emotional stimuli.

Pulse and respiration rates are given in the table below:

	Vostok 3 *Nikolayev* (*min.*)	*Vostok 4* *Popovich* (*min.*)
Pulse frequency 4 hr. before lift-off	72	80
Breathing frequency 4 hr. before lift-off	11	15
Pulse 1 hr. before lift-off	88	100
Pulse 5 min. before lift-off	115	110
Pulse during acceleration to orbit	120	130
Pulse during orbital flight	50–80	50–80
Pulse before re-entry	96–104	85
Breathing frequency before re-entry	11	16

It was also necessary to assess how convenient was the use of the sanitary equipment. The cosmonauts were also expected to assess hygienic conditions in the cabin, including air purity, temperature, humidity, and lighting. The men were also expected to assess the method of taking food and evaluate the food's quality. Meals – more varied than before – were taken four times a day, first breakfast at 0500–0600 hr. (Moscow time), then a second breakfast at 0800–0900 hr., dinner at 1400–1500 hr., and supper at 2000–2100 hr. As well as the tube-packed foods, there were meat cutlets, roast veal, fillet of chicken, pastries, special sweets, miniature loaves, sausages, dragees, and chocolate.

Experiments included floating freely in the cabin while unrestrained by the seat harness. Nikolayev reported spending four periods totalling three and a half hours in this condition and Popovich three periods totalling about three hours. They had to determine how much personal orientation was possible; also what was the most convenient posture to adopt when the muscles were relaxed.

Although at first it was claimed that the cosmonauts experienced no adverse effects of their prolonged exposure to weightlessness, at the 1964 Cospar Symposium it was revealed that post-flight disturbances affecting cardiovascular response persisted for 7–10 days. Professor Vassily Parin, Director of the Institute of Normal and Pathological Physiology of the USSR Academy of Medical Sciences, reported reactions mostly concerned with the heart and blood-vessels. However, he said the central nervous system and metabolism were also affected.

An interesting experiment performed under weightlessness by Popovich concerned the observation of air bubbles in an hermetically sealed flask about two-thirds full of water. When undisturbed, the cosmonaut reported, all the air gathered in the middle of the flask and the water remained round the walls. After the flask was shaken the large bubble split into a multitude of smaller bubbles which gradually merged again into one big bubble. Popovich said he also sprayed water inside the cabin. The water formed small globules which slowly gravitated to the walls and settled on them.

Both cosmonauts described the beauty of the Earth as it passed beneath them. They could easily pick out coastlines, rivers, mountains, and towns; they observed 'sunrises and sunsets', and while over North America they watched thunderstorms in the atmosphere miles below. At the time of the flight the Moon was full and when the craft was in the Earth's shadow the surface appeared 'as a great blanket' against which the cosmonauts could discern the lights of towns.

Yuri Gagarin was on duty at the operations centre near Karaganda with Titov and other cosmonauts. He explained that his task was to maintain direct contact with Nikolayev and Popovich during the countdown and boost periods. He spoke to the men in orbit several times during the 4 days of the operation.

The double landing must have placed a particular load on the ground-control and recovery teams. The flight programme called for landing in Kazakhstan at a latitude of 48°N using an automatic system. First into position was Vostok 3 whose retro-rocket fired at 0924 hr. (Moscow time) on 15 August. Ejecting from the capsule at low altitude Nikolayev parachuted down at 0952 hr. in an area south of Karaganda 48°02′N, 75°45′E. The capsule landed undamaged near by.

Nikolayev's flight was 94 hr. 22 min., during which he completed 64 orbits of the Earth and covered a total distance of 2,639,600 km. (1,639,190 miles).

The retro-rocket of Vostok 4 fired at 0930 hr. (Moscow time); Popovich ejected in the lower atmosphere and touched down south of Karaganda at 48°10′N, 71°51′E. Again the capsule was stated to be undamaged. Flight time was 70 hr. 57 min.; total distance covered, exceeding 48 orbits, was 1,981,050 km. (1,230,230 miles).

With the first Vostok group flight safely completed, it looked obvious that the Soviet space programme was being directed towards the perfection of orbital rendezvous techniques. This view was strengthened by a speech which Yuri Gagarin* gave at the Paris Congress of the International Astronautical Federation in September 1963. He said techniques being worked out in his country involved the assembly and refuelling of spacecraft in orbit. When I saw Gagarin again in 1965 I reminded him of this statement, asking if the Soviet programme was more concerned with a space-station or a Moon-flight. He replied that sending men to the Moon and assembling a space-station were part of the same problem.

The second Soviet group flight strengthened the Western view that orbital rendezvous was Russia's primary objective. Vostok 5

* Gagarin was killed in a flying accident in March 1968 (see page 220).

left the Baikonur cosmodrome on 14 June 1963, at 1459 hr. (Moscow time), entering an orbit inclined at 65° to the Equator, ranging between 112·5 and 146 miles with an initial period of 88·4 min. It carried cosmonaut Lieutenant-Colonel Valery Bykovsky on a flight which lasted 5 days. Bykovsky conducted a series of tests similar to those of his predecessors. Several times he orientated the ship in flight using manual controls; performed various medical tests; observed the Earth, the horizon, the Moon, and the stars; operated the craft's equipment; maintained radio communications with the ground, and floated freely in the cabin.

While the TV camera was operating he demonstrated his new freedom to viewers on Earth. 'I floated up to the portholes of the spacecraft and carried out observations of the ground,' he later explained. 'It is extremely peculiar. The slightest push sends you flying in the opposite direction and, with your eyes closed, you cannot tell what your position actually is.' Apart from the conventional exercises performed by earlier cosmonauts, he did 'power exercises with a rubber strip' which helped to maintain his physical capability to work under weightlessness. He also ate, rested, and slept. He ate four times every 24 hr.; food was the 'ordinary terrestrial kind' and his appetite 'excellent'. He slept well.

Bykovsky said he could easily distinguish rivers, lakes, and oceans. Water in the seas and oceans was of different colours. He described the curvature of the horizon, noting the 'beautiful range of colours, with reddish hues predominating'. On the ground 'roads and towns were visible, towns being particularly clear at night'.

There were amusing moments too. While transmitting a routine report which mentioned 'I had a motion of the bowels', owing to interference ground control received the word as 'stuk' (knocking) instead of 'stul' (bowel motion).

There was uproar in ground control, Bykovsky said: 'I was bombarded with questions. They wanted to know what had gone wrong with the craft. What kind of knocking had I heard? Was it a buzzing or banging? I hurriedly explained that I had merely made use of my sanitary appliance.' In reply, he heard loud laughter: 'My colleagues on the ground were reassured and normal work was resumed.'

Whether or not it had been envisaged that Vostok 6, with the woman cosmonaut Valentina Tereshkova, would be launched the next day to follow the pattern of the first group flight has not been explained. Western observers believe there was a 'technical hold' at the Baikonur cosmodrome which prevented lift-off when Vostok 5 passed overhead on the seventeenth orbit, delaying the launch for another 24 hr. Miss Tereshkova eventually lifted away from the cosmodrome within her Vostok 6 capsule at 1230 hr. (Moscow time) on 16 June. The ship's initial orbit closely matched that of Bykovsky's. Angled at 65° to the Equator, it ranged from 114 to 145 miles with a period of 88·3 min.

Soon after Valentina went into orbit the two craft 'closely passed each other'. The two cosmonauts held their first conversation at about 1300 hr. At 1400 hr. Moscow Television broadcast the first live pictures from Vostok 6. They showed Miss Tereshkova, clad in her spacesuit and helmet, speaking confidently by radio to ground control. She appeared calm and smiling and reported she felt well. By 2200 hr. Vostok 5 was completing its thirty-eighth orbit and Vostok 6 its seventh.

The distance between the two spacecraft varied throughout the flight, from 5 km. (3·1 miles) to several hundred kilometres. Nevertheless, communication between them remained steady.

Telemetered medical data and visual television observations showed that both Bykovsky and Tereshkova were standing up well to orbital flight.

Afterwards the woman cosmonaut gave her reactions at a Moscow Press Conference.

> The flight tasks included work with various equipment in the cabin, the life-support system, and radio communications. I recorded observations in a log-book, on tape and on ciné-film. I endured well the state of weightlessness and quickly adjusted to it. Certainly, it was rather odd to sleep with one's hands hanging in mid-air, but I remembered Gherman Titov's experience and put my hands into my harness while I slept.

> I did not dream. After sleeping I performed physical exercises. I enjoyed my meals. I had a varied diet. True, towards the end I began to want some black bread, potatoes, and onions.

The woman cosmonaut continued:

> I am often asked how I was able to train myself for such an unusual and unfeminine task as piloting a spacecraft. Some people think there is nothing complicated in this, that the automatic equipment aboard the craft functions by itself and that the role of the woman cosmonaut is insignificant. However, a passive role did not suit us. We understood that our role was an active one, most important being manual orientation of the ship and adjustment of the life-support system.
>
> My girl friends and I decided to prepare for the real thing in full measure. This meant determined study, hard work and daily and varied training.

Whereas all the previous cosmonauts had been Air Force jet-pilots, Valentina was a civilian. She was given the title of Junior Lieutenant. Apart from the obvious value of comparing male and female reactions to spaceflight, she was basically a test of any fit and adequately prepared person to go into space. Her background was that of a textile worker who took up parachute jumping as a sport.

Born in a village near Yaroslavl on 6 March 1937, her father was a tractor-driver on a collective farm; he was killed in action in the early days of the Second World War. Valentina started school in Yaroslavl in 1945, when her mother moved there to work at the textile mills.

Valentina herself began work at the age of sixteen, at the Yaroslavl tyre factory, but continued her studies at night school. In 1955 she transferred to the Krasnyi Perekop (Red Canal) textile mill, and at the same time undertook a correspondence course at a technical school. It was largely on her initiative that a parachute-jumping section was set up by the local aero club, and she took part in this sport with great enthusiasm.

After being accepted for training as a cosmonaut in 1962 she learnt to fly various aircraft. It was only then that she began to learn how to control a Vostok. 'I am convinced,' she said, 'that spacecraft leaving for long voyages to other planets will be piloted by our engineers, designers, workers – and textile workers like me.'

Part of the women cosmonauts' training involved flights in a specially adapted Tu-104 transport aircraft. Flown along powered parabolic trajectories this allowed the weightless condition of spaceflight to be simulated for about 30 sec., when the trainees could experiment by floating freely inside the specially padded cabin. Here they learned how to work various equipment under conditions similar to that of orbital flight.

Parachute training was also continued. 'We jumped again and again in different conditions,' Valentina said. 'This gave us a great deal of experience and prepared our morale for the achievement of complex spaceflights.'

She explained how the women worked hard to become familiar with every part of the spacecraft. This was followed by persistent training in the control and operation of the craft.

> It was part of my assignment to guide the craft manually. I switched on the manual control system, noted and recorded gas-pressure of the orientation system and set the stop-watch. The position of the Earth in the porthole was such that it was necessary to orientate the craft in pitch. After this I quickly followed with roll and yaw orientation. I stopped the stop-watch, then observed and recorded the instrument readings. I was glad to see how little fuel I had used. The spacecraft handled well and proved easy to control.

Radio communication during the flight provided much reassurance. 'When Valery and I were talking to each other on short-wave, I had the impression we were sitting in the same room. It seemed the radio waves I was receiving gave me new energy which increased my strength and raised my spirits.'

Valentina said she felt this particularly during her conversations with Mr Khrushchev. 'Yes, Nikita Sergeyevich Khrushchev has become a real radio operator; we followed all the radio communication rules in our conversation. We used the call signals and ended by saying "over".'

The whole flight was full of unforgettable impressions, such as the frequent sunrises and sunsets, and the rapid change in the nature of the terrain below – continents, oceans, seas, clouds, rivers, mountains, towns, and fields.

When she received instructions to return on the third day, all systems worked perfectly; the retro-rocket fired and by the increased g-loading she realized that the capsule had entered the dense layers of the atmosphere. Flames blazed outside the porthole.

As time spent in space increased, so extra safeguards were taken to ensure the continued safety of the cosmonauts. In the case of Valentina Tereshkova and Valery Bykovsky, for example, close observation of activity on the Sun was made before the flight. Professor Anatoli Blagonravov said astronomical observatories were enlisted for these studies. With the help of geophysical rockets a vertical section of the upper atmosphere was studied, which made it possible to obtain, directly, samples of the level of radiation at a given moment. Of the spacecraft, Blagonravov noted that 'substantial improvements were made to the life-support system, providing greater comforts for the cosmonauts'. Whereas previous flights had concentrated on the human organism, it had been possible to include a number of scientific experiments in Vostoks 5 and 6. These included observations of constellations, photography of the Sun and the Earth's disc at sunrise and sunset, as well as observations of the Earth's surface.

Professor Vladimir Yazdovsky, the medical authority, explained that preparations to send a woman into space required special research 'because of the different anatomical and physiological characteristics'. A new system was devised for recording the respiration and cardiac activity, for example, and there were special pre-flight experiments concerning the effects of acceleration and deceleration on the female organism.

Physiological information was telemetered to Earth, but the pulse count was also sent by a special channel of the craft's

Signal transmitter which operated continuously for tracking purposes. Use was made of techniques for studying electrical phenomena of the heart, brain, eyes, and skin, and also seismo-cardiography.

Radiation monitoring during the flight was effected by physical and biological dosimeters. Bykovsky's total radiation dose was 35–40 millirads, and Tereshkova's 25 millirads. Both cosmonauts experienced a certain rise in the heart rate before landing. Bykovsky's pulse rate in orbit varied between 46 and 80 per min. and respiration between 12 and 22 per min. Tereshkova's pulse ranged from 58 to 84 per min. Yazdovsky reported that 'considerable fluctuations in the frequency of heart-beats occurred within short intervals of time'; respiration rate ranged between 16 and 22.

According to Soviet officials, Tereshkova's flight was planned to last 24 hr. However, as her condition remained satisfactory it was decided to allow her mission to continue for 3 days. Yazdovsky explained that 'her sleep during the flight removed the emotional stress and restored the ability to work'. Her pioneer mission, in fact, lasted 71 hr. during which 48 orbits were completed. The total distance covered was 1,970,990 km. (1,224,084 miles).

The return to Earth was made on 19 June at 1120 hr. (Moscow time) some 390 miles north-east of Karaganda on latitude 53°.

Having passed the period of maximum heating, with speed reduced to 493 m.p.h., the hatch cover was automatically fired off at an altitude of 22,966 ft. Valentina ejected in the seat at 21,385 ft. Twenty seconds later, at 13,123 ft altitude, the capsule landing parachute was triggered by the automatic descent control. Valentina herself was still high in the air, swinging lazily beneath her parachute, as her capsule bumped down.

Bykovsky was down 2¾ hr. later, having completed no fewer than 81 orbits of the Earth in a flight lasting over 119 hr.; the total distance covered was 3,325,957 km. (2,065,420 miles). Having ejected from the Vostok capsule he landed at 1406 hr. (Moscow time) some 337 miles north-west of Karaganda, also on latitude 53°.

Although in common with some of her cosmonaut brothers Valentina Tereshkova experienced certain post-flight disturbances in blood-pressure, body chemistry, and nervous responses as determined by measurement of brain rhythms, these effects were stated to have disappeared within 15 days.

A few months after her flight it was announced that Valentina was to marry cosmonaut Andrian Nikolayev. The wedding took place on 3 November 1963.

The following February it was my privilege to meet the remarkable woman whose courageous exploit had thrilled the world. It is the custom of the British Interplanetary Society to make awards for significant 'firsts' in astronautical achievement. Yuri Gagarin received the Society's first gold medallion when he visited London in 1961 and Valentina (now Mrs Nikolayeva-Tereshkova) came specially at the Society's invitation on 5 February 1964.

With other representatives of the Society I was at London Airport to greet the lady following her arrival by Tu-104. One could not fail to be impressed by her modest and gentle manner, and the confident way in which she dealt with questions from the Press. Certainly there were no signs of physical or psychological lassitude; in fact, quite the reverse. Although she was pregnant and her baby had obvious interest for Soviet scientists, there was no special cosseting on that account. She attended all the functions of her strenuous seven-day programme. One evening, though clearly feeling the strain, she stood with the Soviet Ambassador and his wife for nearly three hours welcoming guests at two mammoth Embassy receptions given in her honour.

Naturally, Valentina was proud of her unique achievement. But after receiving the award from Dr L. R. Shepherd, the BIS President, at a special meeting of the Society, she spoke of the 'workers, engineers, technicians, scientists, and all Soviet people' who had made her flight possible. She also acknowledged the large and friendly family of Soviet cosmonauts which contained 'a number of women who are to perform in future many cosmic flights'; she stressed the desire to preserve outer space for peaceful scientific purposes.

Although we were no wiser after her visit about technical features of the Vostok programme, there were many interesting conversations. When I inquired after her husband,* who had been forced by pressure of work to remain in Moscow, Valentina told me he was busy with exams at the Zhukovsky Air Force Engineering Academy. She herself expected to resume her studies after the birth of her baby.

Of her own spaceflight I was interested to know exactly what she had been able to see of the Earth's surface. At the time there was heated debate in America over the detail observations reported by Gordon Cooper from his Mercury capsule. He had described seeing roads, buildings, and smoke from chimneys. Certain experts in visual acuity were frankly incredulous and there was talk of Cooper having suffered hallucinations.

When I put the problem to Valentina she was sceptical of the ability to resolve such fine detail from orbit; but later American observations were to prove the correctness of Cooper's assertions (see page 168).

In the minds of Soviet specialists there had never been any question that Valentina would not have a perfectly healthy baby, though naturally she received special medical care. Just twelve months after she had looked down on the Earth from space, calmly reporting her reactions to scientists on the ground, her baby – a 6 lb 13 oz girl – was born. For some reason news of the birth at the Moscow Gynaecological Institute on 8 June 1964 – reportedly by Caesarian section – was delayed for 48 hr.; but probably this was to allow time for a thorough medical check.

Writing in the March 1966 issue of *Aviatsiya i Kosmonavtika* about her life and future plans, Valentina said she was not only continuing her cosmonaut training but also studying at the Zhukovsky Academy. She was also running her home and looking after her family at the 'Stellar Village' near Moscow where the Soviet men and women cosmonauts receive their training. She denied stories that her daughter, Yelena, had abnormality as a result of her spaceflight.

* Nikolayev replaced Yuri Gagarin as commander of the cosmonaut group in April 1965, being made up to the rank of Colonel.

It should be recalled, too, that there has been one other famous space-mother, the bitch Strelka (Little Arrow) which orbited the Earth 17 times in an unmanned Vostok. Five months after her return to Earth, Strelka had six bright-eyed playful puppies. One, called Pushinka (Fluff), was given by Mr Khrushchev to Mrs Jacqueline Kennedy.

It was at the wedding of Andrian Nikolayev and Valentina Tereshkova in November 1963, that astute Western correspondents noticed among the guests two pioneers of Soviet astronautics. One was Academician Valentin P. Glushko, a fifty-five-year-old combustion engineer who worked on liquid-propellant rocket engines in the 1930s. The other was Sergei P. Korolev, fifty-seven, also prominent in the 1930s in structural aspects of rocket engineering. The connection seemed to confirm these men as prominent in the development of Vostok and its multi-stage launcher. Korolev had previously been identified in a photograph taken at the Baikonur cosmodrome on the morning of Gagarin's flight.

According to Dr G. A. Tokaty, the former Soviet rocket specialist, in 1945 Korolev 'was made responsible for the further development of the German V-2. Later his group designed an ICBM, the successful launching of which was announced on 27 August 1957.' He later became 'one of the chief designers of the Sputnik and Vostok capsule-carrying rockets'.

Korolev's part in the development of Soviet space technology was confirmed shortly after he died on 14 January 1966, when the Soviet Press published an obituary, part of which is reproduced below. It was signed by Leonid Brezhnev, Alexei Kosygin, Nikolai Podgorny, and other Soviet leaders, and also by Academicians Mstislav Keldysh, Mikhail Millionshchikov, Anatoli Blagonravov, and Leonid Sedov.

> Academician Sergei Pavlovich Korolev, a prominent Soviet scientist, a member of the Presidium of the USSR Academy of Sciences, a Communist, twice Hero of Socialist Labour, and a Lenin Prize-winner, died in Moscow on 14 January at the age of fifty-nine. In Sergei Korolev our

country and world science have lost an outstanding scientist in the fields of rocketry and space research, and a distinguished constructor of the first artificial space satellites, which ushered in the era of man's exploration of outer space.

Sergei Pavlovich Korolev was born in the town of Zhitomir on 30 December 1906, the son of a teacher. According to the *Tass* news agency he started work in the aircraft industry in 1927. Three years later he graduated from the Aero-mechanics Department of the Bauman Higher Technical School, having studied without interrupting his work, and in the same year he completed his studies at the Moscow School of Aviation.

After becoming acquainted with Konstantin Tsiolkovsky and with his ideas, Korolev took up space-rocket engineering and became one of its founders. In 1933 he helped to form a group for the purpose of studying rocket flight and this group designed the first experimental rockets. From that time until the end of his life he devoted all his energies to the development of Soviet space-rocket engineering.

Sergei Korolev was a prominent designer of space-rocket systems which were used for launching the first artificial Earth satellites, for carrying the Soviet pennant to the Moon, for a flight round the Moon, and for photographing the side of the Moon which is not seen from the Earth, the news agency reported.

Piloted spacecraft in which man travelled for the first time into outer space and stepped out into space, were designed under Korolev's direction. Sergei Korolev trained large numbers of scientists and engineers who are now working in many space-rocket engineering research institutes and design bureaux.

Korolev lived long enough to see two more Soviet manned spacecraft go into orbit. These were the two Voskhod (Sunrise) vehicles.

As Gagarin had forecast and the flight of Valentina Tereshkova made credible, Voskhod 1 was the first attempt to orbit non-pilot crew-members capable of performing specialist duties. With the late Colonel Vladimir Komarov were a doctor, Boris Yegorov and a scientist, Konstantin Feoktistov.

Yegorov's experience was particularly important as it allowed medical experience to be applied, for the first time, directly to conditions as they existed in orbit.

The first Voskhod entered an orbit, inclined at 64°54′ to the Equator, between 100 and 255 miles from the surface.

Few details of the 5,320-kg. (11,731-lb) spacecraft were given following the launching which occurred at 7 hr. 30 min. 1 sec. on 12 October 1964. Such photographs as were officially released suggested similar external dimensions to the Vostok. The launch vehicle was clearly an uprated version of the standard multi-stage rocket. The re-entry capsule had been re-designed to provide accommodation for three men seated side-by-side; apparently the centre position was slightly forward. Although Gagarin later told me that Voskhod was an entirely new design, it clearly owed much to existing Soviet components and techniques.

However, there were no ejection seats and the crew dispensed with pressure suits. A pressure of 1·1 atmospheres was maintained in the cabin. The men wore silver-grey woollen suits with blue jackets and white helmets with headphones. Pressure suits for emergency use were stowed in a locker. New systems incorporated in the vehicle included two sets of retro-rockets, main and reserve, allowing the ship to enter a higher orbit with greater margins of safety. Television cameras permitted 'transmission not only from the cabin but also of the ship's environment', and an attitude control system included 'ion-plotters of the direction of the ship's velocity vector'. The Russians have since admitted that Voskhod 1 differed from Vostok in having a soft-landing system, a reserve solid-propellant retro-rocket, new instruments, an additional orientation system and improved TV and radio equipment.

The ship had two ultra-short-wave and short-wave radio-telephone channels, one telegraph system, and a radio receiver operating on short and medium wavebands. There were also 'innovations' in the method of direction finding.

During the flight Konstantin Feoktistov described luminous particles which at times could be observed outside the portholes. He thought they could be dust particles from the body of the ship. The crew gauged the particles' distance against the antenna;

it proved to be about 3 ft. In space such particles could be expected to travel with the craft but, Feoktistov remarked, 'we observed a slow relative movement'. They were between 5 and 50 microns in size and like terrestrial dust particles, 'they shone in the sunlight and were visible for a few dozen seconds'.

As the cosmonauts aboard Voskhod 1 had no means of ejecting from the re-entry capsule, they had to remain inside for the touchdown. This occurred some 312 km. (194 miles) north-east of Kustanai in Kazakhstan. The landing parachute opened at 16,000 ft when the speed was about 720 ft/sec. In order to soften the touchdown the vehicle had a system of braking rockets which ignited just before the capsule, hanging beneath its parachute, made contact with the ground. Sixteen orbits were completed in a 415,936-mile flight lasting 24 hr. 17 min. 3 sec. The capsule 'was intact and could be used again'.

Yet another milestone in space history was reached at 1000 hr. (Moscow time) on 18 March 1965, when Voskhod 2 lifted off from the Baikonur cosmodrome. Inside were two men, Colonel Pavel Belyaev and Lieutenant-Colonel Alexei Leonov.

Orbiting between 107 and 308 miles above the Earth, at the standard 65° inclination, they travelled higher than anyone had gone before. The mission had one major and dramatic objective: to allow the spacesuited Leonov to emerge from the craft in orbit by means of an airlock.

Records of the mission submitted to the International Aeronautical Federation in Paris stated that Leonov spent a total of 23 min. 41 sec. outside the cabin. Total weight of Voskhod 2, without the rocket's final stage, was 12,529 lb, total distance travelled 445,420 miles, and the flight duration 26 hr. 2 min. 17 sec.

Television pictures of Leonov leaving the ship suggested that the spacecraft was a 'stretched' version of the Vostok. This was later confirmed by photographs of the nose-cone of the launch vehicle in a preparation building.

Leonov was seen emerging from a hatch opening at the end of an airlock which appeared to come from the side of the space capsule. Side-by-side seating enabled him to leave the cabin from the left side, the airlock itself apparently being inflatable. Once the cosmonaut was inside, the cabin hatch was closed while Leonov, secure in his suit, waited for air to be pumped from the airlock chamber. When vacuum-equivalent conditions were obtained, the outer airlock doors were opened and he climbed out in space. Without the airlock, it would have been necessary to depressurize the entire cabin before egress as indeed was done in the case of the American Gemini spacecraft.

Few details were forthcoming on the design of the extra-vehicular pressure suit. A Soviet medical specialist, Vladimir Krichagin, described it as 'a miniature hermetic cabin which consists of a metal helmet with a transparent visor, a multi-layer hermetic suit, gloves, and specially designed footwear'. The suit had its own power circuitry feeding communications, and an arrangement of sensors giving data on physiological functions.

Internal air pressure, Krichagin said, had to be at least 0·4 atmosphere. Even then the suit inflated considerably and it was tiring to work inside without articulated joints.

Before going into space, nitrogen had to be eliminated from the cosmonaut's organism; prolonged respiration in pure oxygen literally washes nitrogen out of the tissues of the body and then the pressure can be safely reduced. In space there must be a steady supply of pure oxygen for the cosmonaut. It should be borne in mind that his body too, has to 'breathe'; every hour a human being gives off up to 300 kilo-calories of heat to the surrounding media. Left uncontrolled the cosmonaut's body temperature would therefore rise and he could suffer heat-stroke. Hence Leonov's spacesuit had a special air-conditioning system through which air circulated at room temperature; this air carried away excess heat of the body and moisture exuded from the skin.

To protect the cosmonaut from solar heat and from cold in the shadow of the Earth or the spacecraft, the suit was covered by a strong layer of thermal insulation and coloured white. Air for ventilation and oxygen for respiration, Krichagin said, could be taken from the spacecraft's supply or from cylinders mounted on the cosmonaut's back. The air used was rejected into the environment.

Leonov emerged from the airlock as the spacecraft entered its second orbit. When I saw him at the Space Congress in Athens in 1965, he confirmed that he had remained outside for about 10 min.; another 10 min. was spent inside the airlock – a total of 20 min. in vacuum. He had been connected with the spacecraft by a lifeline which included 'a telephone cable and telemetric wires'. No oxygen reached him through the tether; he was dependent entirely on the life-support pack on his back. Leonov regulated his own suit pressure. There were two settings. Before he left the cabin it was 0·27 atmosphere; when he stepped out into space it was 0·4 atmosphere and he reset it to 0·2$\frac{1}{4}$ before climbing back. Belyaev wore a similar extra-vehicular suit.

Leonov fixed a television camera to a bracket on the edge of the open airlock before climbing out. There was a second camera outside the ship. The externally mounted cameras showed Leonov launching himself gently into space and performing gyrations at a distance of some 15 ft. His easy movements about his centre of mass and the way he was able to control his tumbling action by arm motions gave added confidence that man would be able to work effectively in space if necessary on the outside of his spacecraft.

Of the future, Academician M. V. Keldysh said: 'Use of an airlock instead of cabin depressurization would allow groups of cosmonauts to emerge from a ship in orbit and transfer from one craft to another.'

Yet still this flight did not demonstrate the ability of Voskhod to manœuvre from one orbit to another, essential for perfection of the technique of orbital rendezvous. However, at a Moscow Press Conference, Keldysh remarked that manœuvres were possible with this craft similar to those performed with earlier, unmanned, Polyöt sputniks. It could have remained in space for a month.

Despite the apparent success of the Leonov experiment, the mission did not end without difficulty. It had been planned to bring the craft back to a landing in Central Asia on the seventeenth orbit. However, a fault in the solar sensor of the attitude control system prevented firing of the retro-rocket and the crew were forced into making another complete orbit of the Earth. Re-entry was manually controlled by Belyaev. Instead of returning in the lower latitudes of Kazakhstan, the trajectory was displaced westward due to the Earth's rotation and the capsule came down perilously in a snow-covered forest near Perm in the Urals. This was some 750 miles north-east of Moscow. A communications aerial burned off during re-entry resulting in loss of contact. Nevertheless, both cosmonauts were reported to be in good health following the flight.

Belyaev later explained that they were out of the capsule within 5 min. However, it was two and a half hours before the first rescue helicopter arrived and it was extremely cold. They were taken to Perm by air the next day. Meanwhile their capsule, stated to be undamaged, was air-lifted back to the Baikonur cosmodrome.

Leonov nearly missed his chance of becoming the world's first 'space-walker'. At the end of 1963, he and his wife were returning by car to the 'Stellar Village' when the driver misjudged a corner. The vehicle left the road, ending up through the ice of a pond. Leonov struggled out rescuing his wife and the driver from under the water.

There seems little doubt that features of the Voskhod and its successors were developed in secret as part of the Cosmos programme. From a comparison of orbital parameters, it is possible to conclude that Cosmos 47 – recovered after 1 day – was a trial run for Voskhod 1 introduced 6 days later.

Few details of recoverable Cosmos satellites have been released by the Soviet authorities, and US sources believe a number have been launched for reconnaissance purposes.

Such vehicles have been observed to return their capsules over the Soviet Union after a few days. However, many of the Cosmos breed – both recoverable and non-recoverable – have legitimate scientific tasks, not least of which has been the measurement of air drag and radiation at different altitudes.

Cosmos 110, launched on 22 February 1966, helped to prepare the way for the next series of Soviet manned spacecraft. It was devoted to a thorough investigation of the working of the heart and the entire circulatory system in the space environment. On board were two dogs Veterok and Ugolek, in separate pressurized containers. The first was the principal experimental animal, the second the control animal.

The orbit achieved, inclined at 51·85° to the Equator, ranged between 118 and 548 miles, taking the craft into the region of the inner Van Allen radiation belt.

The re-entry capsule of Cosmos 110 was recovered on 16 March after a flight of nearly 22 days, with its dog-passengers 'alive and well'. Dr Boris Yegorov commented: 'We have got an answer to two main questions: Whether changes occurring in the organism represent adaptation to weightlessness; and, if such adaptation is possible, how far the organism will be able to compensate for the increased load on returning to Earth.'

Clad in special suits, which secured them in the capsule, they could nevertheless assume different positions; the suits also served to connect sensors and tubes to the animals for bio-medical studies and food delivery.

Paste-like food was introduced directly into the animals' stomachs from plastic containers by pneumatic action through artificially made openings (gastrostoms). The food, which had been tried out in advance, included meat, potatoes, flour, vitamins, and water.

Air-conditioning and regeneration systems were described as an improved modification of identical systems used during the flights of Soviet spacecraft with dog-passengers in August 1960 and March 1961.

For studying neuro-reflectory regulation of the dogs' cardiovascular system, the capsule was equipped to measure arterial pressure, record bio-currents of the heart with the aid of ingrown electrodes, the pulse of the carotid artery, the mechanical working of the heart, and breathing. In addition electrodes were ingrown into peripheral nerves to gauge the activity of the central formations of the brain regulating the vascular tonus. Bio-medical data were telemetered to Earth to ensure timely medical control over the state of the dogs' health.

On television Dr Yegorov demonstrated the 'canine doubles' of Veterok and Ugolek. They were wearing space-jackets to which sensor leads and feeding devices were attached. He pointed out the tube grafted into a dog's aorta by means of which arterial pressure is measured. After the experiment, he said the tube would be painlessly removed without any ill-effect to the dog's vital functions. Dr Yegorov also drew attention to an orifice plate through which food is pneumatically forced into the dog's stomach. Viewers saw a plastic bag with this food weighing 600–700 grammes. One bag had enough food for 24 hr.

The long delay in Soviet manned launchings which followed the flight of Voskhod 2, therefore, did not represent a period of inactivity. Much of the intervening development was cloaked in secrecy, but clearly attention was given to perfecting an extra-vehicular spacesuit and a more reliable airlock system alongside the development of the larger spacecraft and multi-stage booster.

It is instructive to return to the comments of Yuri Gagarin at the 1963 IAF Congress. 'Techniques being developed in my country', he said, 'involve the assembly of components of space-craft in Earth-orbit and the introduction of propellant.' He said this procedure was being adopted because it was not possible to launch vehicles of several scores of tons directly to the Moon.

Then, in October, came Nikita Khrushchev's famous remark about watching the Americans reach the Moon. 'We will see how they fly there, and how they land there . . . and, most important, how they will take off and return.' Was the pace becoming too hot, or had America misjudged Soviet intentions of being first on the Moon?

Although the group flights performed by Nikolayev and Popovich in 1962 and Bykovsky and Tereshkova in 1963 were impressive, particularly in launch timing and orbital injection, they did not amount to orbital rendezvous. Gagarin himself pointed out that although it required little extra propellant to achieve rendezvous when the distance separating spacecraft was down to a few miles, there were still difficult problems affecting 'communications, optics, and manœuvre'. The Americans, of course, solved these problems in the Gemini programme with the help of radar.

Gagarin's remarks in Paris were given added point in an article he wrote commemorating the sixth anniversary of Sputnik 1. 'It may, of course, be too bold of me to conclude that interplanetary travel will be a fact within a few years. Preparations for these flights will call for a still greater effort including many more Earth-orbital flights.'

He stressed again the key nature of Earth-orbit rendezvous. When these experiments are finalized, 'we shall be able to assemble spacecraft of any size directly in flight and the refuelling problem, which is so important for protracted space journeys, will also be solved'.

When at last rumours of an impending new Soviet spaceflight began in Moscow towards the end of April 1967, they were linked with the prospect of two manned vehicles achieving a space rendezvous. In the preceding months an increased tempo of launchings in the Cosmos programme had appeared to indicate the end of unmanned testing of a spacecraft, larger than Voskhod, which would orbit at an inclination of 51–52° to the Equator.

We were not long to be kept in doubt. At 0335 hr. (Moscow time) on 23 April, Soyuz (Union) 1 ascended from the Baikonur cosmodrome with a single occupant, Colonel Vladimir Komarov, the man who had previously commanded the three-man Voskhod 1. The orbit, inclined at 51°40′ to the Equator, ranged between 125 and 139 miles. Reliable two-way radio contact with the spacecraft was reported on frequencies of 15·008, 18·035 and 20·008 Mc/s.

According to *Tass* the mission involved testing a new piloted spacecraft; checking the craft's systems and elements in conditions of spaceflight; conducting extended scientific and physical-technical experiments and studies in conditions of spaceflight, and continuing medical and biological studies and studies of the influence of various factors of spaceflight on the human organism.

When Soyuz 1 completed its fifth orbit at 1000 hr. (Moscow time), Komarov reported that the flight programme was being fulfilled successfully and that he was well. Cabin temperature as indicated by telemetry was 16°C and cabin pressure 750 mm. of mercury – both normal. Radio communication was stable.

Between 1320 and 2120 hr. the cosmonaut rested out of radio contact with the command station. By 2230 hr. the craft had completed its thirteenth revolution with Komarov reporting that all was going to programme. At 0450 hr. on 24 April Komarov again reported his excellent physical condition which was confirmed by telemetered data. Cabin temperature then was 17·5°C and pressure 800 mm. of mercury. A number of set experiments had been completed. Moscow fell silent on the flight after this until 1723 hr. when the shock announcement came of Komarov's death.

An official communiqué explained that the cosmonaut was in touch with ground stations on the nineteenth orbit as he began his re-entry over Africa reporting, 'In good health, systems functioning well.' The time was about 0600 hr. (Moscow time). After the capsule had successfully penetrated the atmosphere the shroud lines of its main landing parachute became entangled at a height of some 7 km. (4·3 miles) and Komarov plunged to his death over the Soviet Union.

However, there were unofficial reports that during its last orbits Soyuz had been rolling out of control because of a failure in the attitude control system.*

* On 5 May 1975, eight years after the fatal accident, a National Security Agency technician manning a USAF listening post in Turkey, was reported as saying (*Daily Telegraph*, 6 May 1975) that the USA had definite proof that Komarov was in trouble long before he fell from the skies. His troubles were said to have begun during the last 12 hr. of the mission with the ship rolling out of alignment every so often. After 5 hr. trying every mechanical possibility, the spinning got worse. Soviet ground control made desperate efforts to control the semi-automatic spacecraft and Komarov himself, ill from the effects of disorientation, had fought to operate its controls. Before time ran out for the cosmonaut, emotional conversations passed between Komarov and his wife, and Komarov and Premier Kosygin.

Although the retro-rocket must have fired at about the right angle to initiate re-entry, the command module (having presumably separated from its service and orbital modules) must have been badly seared by heat as, still spinning, it made its re-entry giving no chance for the parachute to operate in a correct manner. The capsule plunged into hilly country outside Orenburg. A memorial now marks the spot.

Two days after the tragedy the ashes of a brave pioneer of the space age were buried in the Kremlin wall alongside those of Soviet statesmen.

There was no question that the accident would deflect the path of the Soviet programme. A State Commission was immediately set up to inquire into every aspect of the failure and ensure that further spaceflights could proceed in safety. It was the year of the fiftieth anniversary of the Communist Revolution and Russia was on the brink of important space developments.

When flight testing resumed the craft was fitted with a back-up parachute system and other safety devices. Soyuz 2, launched unmanned on 25 October 1968, achieved an orbit of 170–210 km. (106–130 miles) inclined at 51·6° to the Equator. Twenty-four hours later came the launch of Soyuz 3 with a single occupant Colonel (now Major-General) Georgi Beregovoi, at forty-seven the oldest man in space. His orbit was 177–203 km. (110–126 miles) allowing the two spacecraft to make a close approach soon after the launching.

These tests appear to have been satisfactory. The command module of Soyuz 2 landed in the Soviet Union after 2·95 days. Soyuz 3, which touched down on 30 October, completed 61 orbits in a flight lasting 94 hr. 51 min.

What *Tass* described as the world's first experimental space-station was established for 4 hr. 35 min. on 16 January 1969 when two manned spacecraft, Soyuz 4 and 5, were successfully docked in Earth-orbit. Each craft had a spherical orbital compartment in addition to the re-entry module for scientific experiments giving a combined habitable volume of 18 m.^{3} (635·7 ft^3). The station's orbit ranged between 209 and 250 km. (129·7 and 155 miles) inclined at 51°40′ to the Equator.

Soyuz 5, which became the passive vehicle during rendezvous and docking, contained three cosmonauts: Lieutenant-Colonel Boris Volynov, Aleksei Yeliseyev, and Lieutenant-Colonel Yevgeny Khrunov. The active craft Soyuz 4 was launched with a single occupant, Lieutenant-Colonel Vladimir Shatalov, who took over manual control when the distance between the two vehicles was down to 100 m. (328 ft). The subsequent nose-to-nose docking achieved mechanical and electrical interconnection, and telephone communications were immediately established between all four habitable compartments of the composite station.

Still more spectacular events were to follow. During the thirty-fifth orbit (of Soyuz 4) Soyuz 5 cosmonauts Khrunov and Yeliseyev donned spacesuits, the orbital compartments of both craft were sealed and depressurized, and the two men emerged through a hatch into outer space. Their work outside the station included installation and dismantling of ciné-camera supports, handrails, and TV equipment. The men also performed various movements simulating more complicated assembly tasks. After an hour, they entered the hatch of Soyuz 4 and, when the orbital compartment had been sealed and repressurized, removed their suits. The interconnecting hatch to the command module was then opened and they took their places beside cosmonaut Shatalov.

When the docking exercise was completed the craft were separated and their re-entry modules safely recovered. Thus ended the first transfer of cosmonauts from one spacecraft to another in orbit, with three men returning in the vehicle that originally had contained only one.

Academician Mstislav Keldysh, president of the USSR Academy of Sciences, linked the experiment with the broad objectives of the Soviet space effort which included 'the investigation of near-Earth space and planets; the assembly of big, constantly operating orbital stations; interplanetary flights, and advances in radio, TV and other branches of science and technology'.

Thirty-one separate manœuvres were carried out during the group flight of Soyuz 6, 7 and 8. The craft changed their orbits on a number of occasions, approached each other and then separated. Observers in Britain saw the three craft in orbit after dusk. Soyuz 7 and 8 were close together, possibly 8 miles apart. Soyuz 6 was approximately 300 miles behind.

According to *Tass*, only certain manœuvres were made by command from ground control. Ship commanders carried out mutual manœuvring, approach and separation using manual controls. Docking was not part of the programme. Crews carried out various experiments and space manœuvres independently.

One of the most important in-flight experiments concerned navigation. A Soviet control engineer explained: 'As distinct from air and sea navigation, space navigation does not answer the question of where the craft is at a given moment; while the question is being asked the vehicle will have covered some 40 km. in orbit.'

In Soyuz 8, Aleksei Yeliseyev tried out a new sextant which helps a cosmonaut determine the elements of his orbit independently without help from a ground station. He was able to check the results with the help of an on-board computer whilst, simultaneously, the orbital parameters were obtained by ground stations using a highly accurate radio technique.

Another Soviet specialist spoke of the wide range of scientific and technological tasks accomplished during the group flight. Manœuvres – to within a few hundred metres – were conducted mainly through manual control but with the help of automatic systems on board. The craft were also landed by the cosmonauts working manual controls. 'These experiments', it was stated, 'provided important data for designing new systems of autonomous control.'

Three methods of automatic welding (plasma, electron beam and arc) were tested by remote means within the evacuated orbital compartment of Soyuz 6. *Novosti* said the automatic welding was performed with a Vulkan experimental unit linked telemetrically with the command module by cable. It comprised two sections. One contained various instruments and power sources, control instruments, measuring and converter devices, and communication and automation equipment. They were kept in a compartment filled with nitrogen.

Development of this equipment began in a ground laboratory within a conventional vacuum chamber; then small devices, special vacuum chambers and vacuum pumps were mounted in a test aircraft where conditions of weightlessness and vacuum close to those of orbital space were created and conditions of depressurizing a spacecraft simulated.

Academician Boris Paton, director of the Kiev Institute of Welding, said the results of preparatory work have already

THE SOYUZ PROGRAMME

Spacecraft	Date	Crew	Notes
Soyuz 1	23–24 Apr. 1967	Vladimir M. Komarov	Cosmonaut killed in re-entry accident. Flight lasted 26 hr. 40 min.
Soyuz 3	26–30 Oct. 1968	Georgi Beregovoi	Manœuvred near unmanned Soyuz 2. Flight lasted 94 hr. 51 min.
Soyuz 4	14–17 Jan. 1969	Vladimir Shatalov	Soyuz 4 and 5 docked and transferred two cosmonauts from Soyuz 5 to Soyuz 4.
Soyuz 5	15–18 Jan. 1969	Boris Volynov, Aleksei Yeliseyev, Yevgeny Khrunov	Soyuz 4 flight lasted 71 hr. 22 min.; Soyuz 5 72 hr. 40 min.
Soyuz 6	11–16 Oct. 1969	Georgy Shonin, Valery Kubasov	Soyuz 6, 7 and 8 operated as a group flight without actually docking. Each conducted
Soyuz 7	12–17 Oct. 1969	Anatoly Filipchenko, Vladislav Volkov, Viktor Gorbatko	certain experiments, including welding in Soyuz 6, and Earth and celestial observation.
Soyuz 8	13–18 Oct. 1969	Vladimir Shatalov, Aleksei Yeliseyev	Flight times were 118 hr. 21 min.; 118 hr. 43 min. and 118 hr. 51 min. respectively.
Soyuz 9	1–19 June 1970	Andrian G. Nikolayev, Vitaly Sevastyanov	Longest-duration manned spaceflight to date. Mission included experiments in space navigation and Earth and celestial observations 'of interest to the national economy'. Mission lasted 424 hr. 59 min., completing 286 orbits.
Soyuz 10	22–24 Apr. 1971	Vladimir Shatalov, Aleksei Yeliseyev, Nikolai Rukavishnikov	Docked with Salyut 1, but crew did not board space-station launched 19 April. Flight lasted 47 hr. 46 min.
Soyuz 11	6–30 June 1971	Georgi Dobrovolsky, Vladislav Volkov, Vicktor Patseyev	Docked with Salyut 1 for 22 days, but crew perished as they returned to Earth, when faulty valve allowed air to escape from cabin. Flight lasted 570 hr. 22 min.
Soyuz 12	27–29 Sept. 1973	Vasily Lazarev, Oleg Makarov	Test of improved Soyuz for Salyut ferry missions, with chemical batteries replacing solar cell arrays. Flight lasted 47 hr. 16 min.
Soyuz 13	18–26 Dec. 1973	Pyotr Klimuk, Valentin Lebedev	Test wide range of experiments related to Salyut programme – astrophysical, biological, Earth resources. Flight lasted 188 hr. 55 min.
Soyuz 14	3–19 July 1974	Pavel Popovich, Yuri Artyukhin	Docked with Salyut 3 for 14 days 17 hr. 33 min. Programme included medical, biological, atmosphere physics and Earth-resource experiments. Omitted solar arrays; chemical batteries rechargeable from solar arrays of Salyut.
Soyuz 15	26–28 Aug. 1974	Gennardy Sarafanov, Lev Demin	Rendezvoused with Salyut 3 on 27 August but failed to dock because of fault in automatic control system. Launched and recovered in darkness. Flight lasted 2 hr. 12 min.
Soyuz 16	2–8 Dec. 1974	Anatoly Filipchenko, Nikolai Rukavishnikov	ASTP flight test. Tested docking system of Soyuz and reduction of cabin pressure as required for joint docking experiment with Apollo CSM in July 1975. Flight lasted 143 hr. 4 min.

THE SOYUZ PROGRAMME

Spacecraft	Date	Crew	Notes
Soyuz 17	11 Jan.–9 Feb. 1975	Alexei Gubarev, Georgi Grechko	Docked with Salyut 4 for 28 days 5 hr. 8 min. Programme included checks on Salyut's improved design and systems, and research into space, the Earth's atmosphere and surface, and medical and biological tests. Omitted solar cell arrays. Flight lasted 709 hr. 20 min.
Soyuz 'anomaly'	5 Apr. 1975	Vasily Lazarev, Oleg Makarov	Third stage malfunction of launch vehicle. Cosmonauts – attempting to reach Salyut 4 – make emergency recovery downrange from cosmodrome in Western Siberia.
Soyuz 18	24 May–26 July 1975	Pyotr Klimuk, Vitaly Sevastyanov	Docked with Salyut 4 to continue experiments. Omitted solar cell panels. Flight lasted 1,511 hr. 20 min.
Soyuz 19	15–21 July 1975	Alexei Leonov, Valery Kubasov	ASTP mission. Docked with Apollo CSM/DM for about 2 days for crew exchanges and joint experiments. Flight lasted 142 hr. 31 min.

Additional details of Soyuz ferry missions to space-stations will be found in Chapter 6 and the companion volume *Frontiers of Space*.

proved beneficial in general industry. Arc welding technology, for example, was improved as a result of the space requirements and highly efficient small devices designed for joining metals by electron beam and compressed arc.

In space, the research would lead to equipment for the assembly and repair of orbital stations and antennae for radio-astronomy, according to an anonymous Soviet welding expert.

Another interesting facet of the group flight was the emphasis given to Earth resources survey. Soyuz 7, for instance, was stated to have photographed the Earth extensively in a way 'that could benefit the national economy'. Experiments were carried out 'to study geological areas to assess the possibility of finding reserves of mineral raw materials'.

Distribution of snow cover and ice was determined.

There were also experiments to determine the reflective properties of forests, deserts and other areas of the Earth's surface.

The group flight also demonstrated Soviet ability to bring down a succession of manned spacecraft on land, a necessary prelude to space-station assembly. This was in marked contrast to the huge recovery operations that had to be mounted at sea each time an American spacecraft returned to Earth.

It was emphasized that all craft made use of aerodynamic lift on re-entry (as did the US Gemini and Apollo) and that g-forces were small. The three craft landed as follows: Soyuz 6 at 180 km. north-west of Karaganda at 1252 hr. (Moscow time) on 16 October; Soyuz 7 at 155 km. north-west of Karaganda at 1226 hr. (Moscow time) on 17 October and Soyuz 8 at 145 km. north of Karaganda at 1210 hr. (Moscow time) on 18 October.

Soyuz 9, launched on 1 June 1970, continued the same kind of experiments as Soyuz 7, including Earth-resources observations, but in addition the cosmonauts were heavily instrumented for bio-medical research in an orbital flight that was destined to break all previous records. The spacecraft completed 286 orbits in a mission lasting 424 hr. 59 min.

The landing was made in a ploughed field near the village of Intumak some 75 km. west of Karaganda in the presence of four helicopters of the rescue team including medical specialists.

SOVIET SALYUT SPACE STATIONS

Salyut	Launch Date	Orbit* (km)	Inclination to Equator (deg)	Lifetime (days)	Spacecraft visits
1	19 Apr. 1971	200–210 After 9th day 251–271	51·56	175	Soyuz 10, Soyuz 11
2	3 Apr. 1973	207–248 After 5th day 257–278	51·56	55·11	None; station broke up in orbit
3	24 June 1974	213–253 After 4th day 266–269	51·58	214	Soyuz 14, Soyuz 15
4	26 Dec. 1974	212–251 After 4th day 276–341	51·57	N.A.	Soyuz 17, Soyuz 18

* Figures supplied by R.A.E., Farnborough. N.A.=Not available.

An official statement explained the method of recovery. First to open was a small drogue parachute, followed by a braking parachute, the function of which is to reduce speed sufficiently for the main parachute to open. There was also a back-up system embodied since the Soyuz 1 landing accident. If the primary system failed, 'the entire parachute system would have been immediately discarded and the reserve one would have replaced it'.

CHAPTER TWO

BRIDGE TO THE MOON

Future historians may well refer to our time as 'the era of the lunar journey'. Spawned by the missile and space rivalry that developed between Russia and America in the 1950s, its foundations in the United States were laid in 1958 when Congress passed the National Aeronautics and Space Act calling for leadership in space.

Project Apollo was sealed on 25 May 1961, by the late President Kennedy. Addressing a joint session of Congress, he said: 'I believe that this nation should commit itself to achieving the goal, before the decade is out, of landing a man on the Moon and returning him safely to Earth. No single space project in this period will be so difficult or so expensive to accomplish.'

In setting this ambitious goal, the Kennedy Administration recognized the necessity of a challenging objective. If men could be landed on the Moon, the technology involved would provide a solid base from which other space ventures could be sprung, either military or peaceful according to need.

But much of the spadework had been done earlier under the Eisenhower Administration. In October 1958, the National Aeronautics and Space Administration (NASA) had come officially into being on the basis of a reorganized National Advisory Committee on Aeronautics (NACA). Under administrator Dr T. Keith Glennan, a primary task had been to define the first American manned spacecraft, drawing upon proposals previously reviewed by NACA. The Space Task Group, initially an informal team at Langley Field, Virginia, under the direction of Dr Robert R. Gilruth, subsequently became the basis of the Manned Spacecraft Center* at Houston, Texas. Contracts for the first US manned space capsule were placed by NASA in 1959. The name Mercury was chosen symbolic of the winged messenger of the gods of Roman mythology.

McDonnell Aircraft Corporation, the prime contractor,

* Now the Johnson Space Center.

delivered the first of an initial batch of twenty-four development capsules only fourteen months after receiving the order. These 'boiler-plate' vehicles were used for various purposes, including the test of the rocket escape system designed to pull the capsule away from its launching rocket in the event of a catastrophic failure either on the launch pad or following lift-off. These tests were assisted by solid-propellant rockets known as Little Joe. Unlike the Russian Vostok capsule the Mercury astronaut was not provided with an ejection seat. However, once the escape rocket had separated the capsule from its booster, standard recovery procedures would release the drogue and landing parachutes for a normal recovery in the sea.

Out of 508 suitably qualified jet-pilots who applied as Mercury astronauts, thirty-two were selected for preliminary physical and mental tests. On 2 April 1959, the names of the seven successful candidates were announced by NASA. They were: Lieutenant-Commander M. Scott Carpenter; Major L. Gordon Cooper, Jr; Lieutenant-Colonel John H. Glenn, Jr; Captain Virgil I. Grissom; Commander Alan B. Shepard, Jr; Commander Walter M. Schirra, Jr, and Major Donald K. Slayton. Slayton subsequently was found to have a slight heart irregularity and was removed from the roster of active astronauts. However, his function in the Mercury programme should not be minimized. He served as co-ordinator for astronaut activities, and maintained overall supervision of astronaut duties. (Another nine astronauts were chosen by NASA in September 1962 to participate in the subsequent Gemini programme.)

The first major firing of a boiler-plate Mercury capsule on an Atlas-D occurred on 9 September 1959. The test was sub-orbital and meant to subject the capsule to severe heating and airloads in ballistic flight. The test was successful and the capsule was recovered intact.

The programme did not proceed without its frustrations. Another key test was scheduled for 29 July 1960, when a production capsule would be launched unmanned from an Atlas booster on a sub-orbital flight. After travelling 1,500 miles over the Atlantic from Cape Canaveral, it was meant to reach its zenith some 110 miles above the Earth; the capsule would plunge back into the atmosphere at 19,000 ft/sec. enduring decelerative forces up to 16·5 g. As there was no test-subject aboard the escape system was omitted.

As it happened the booster exploded catastrophically about a minute after leaving the pad. The instrumented capsule, blown clear of the Atlas, was intact as it hit the water; NASA salvaged the wreckage.

Meanwhile, another smaller rocket called Redstone, developed by Dr Wernher von Braun's Army Group, was being pressed into service in order to gain further experience with Mercury capsules in sub-orbital flight. The rocket had the capability of boosting a capsule over a ballistic trajectory of some 200 miles reaching its peak 125 miles into space.

By this time the seven volunteer astronauts – all seasoned military test-pilots – were busily preparing for the day when they would occupy the 'hot-seat' on top of a rocket on its way into space. Much preliminary work had already been done in America with the celebrated X-series of rocket-aircraft culminating in the X-15. Bio-medical experiments had been made with animal and human test-subjects in centrifuges and rocket-sleds. As early as 1952, it was disclosed that a number of rhesus and cebus monkeys and white mice had, on occasions, been rocketed to altitudes of up to 80 miles in V-2 and Aerobee rockets.

In these pioneer experiments by the US Air Force the monkeys had been anaesthetized before take-off, a 'luxury' subsequently denied to later experimental animals. Strapped down by nylon netting on sponge-rubber beds, they were placed inside pressurized capsules 3 ft long by 15 in. diameter.

Each animal had a face-mask through which it received a recirculated supply of oxygen. Instruments attached to their bodies allowed measurement of blood-pressure, heart action, pulse, and respiration, and this information was telemetered to the ground.

Data received indicated that the monkeys were not seriously disturbed by the actual flight. However, although the capsules were successfully ejected from the rockets, only one reached the ground alive because of failure of the landing parachutes. Ironically, the one monkey that did reach the ground safely died of heat-stroke in the New Mexico desert a short time before it was located.

An Aerobee launched from Holloman Air Force Base in 1952 was particularly successful in showing the reactions of mammals under weightless conditions. Inside the nose-cone were two monkeys and two white mice. The mice travelled in a glass-walled chamber with a ciné-camera to record their behaviour. One mouse had part of the balance mechanism of the inner ear removed.

When the nose-cone was recovered and the film developed, the mice were found to have acted normally up to the time propulsion ceased, but during the 3-min. period of zero gravity they were seen in different attitudes threshing around and oddly suspended between floor and ceiling.

Dr J. P. Henry of the Wright Field Aeromedical Laboratory, under whose aegis the experiments were made, said the mice were apparently as much at ease when inverted as when upright; they merely lost the vertical reference of gravity and assumed whatever posture was convenient. The mouse that had been deprived of its balance mechanism remained curled up in a corner of the capsule, seemingly unaware of its weightlessness.

The mice were returned to the laboratory where they were reported to be lively and breeding freely. Together with the two small monkeys, housed in separate compartments, they had endured a brief initial acceleration of 15 g, lasting less than 1 sec., and a longer force of 3 to 4 g lasting for 45 sec.

Dr Henry's conclusions, nearly a decade before the first man rocketed into space, were extraordinarily perceptive; he said pilots would have no difficulty in performing all actions necessary to control a vehicle in a weightless state.

Primates were used in a number of subsequent space experiments. Particularly successful, on 28 May 1959, was the ballistic flight of monkeys Able and Baker in the re-entry cone of a Jupiter IRBM launched on test down the Atlantic Missile Range. They were recovered after reaching a peak altitude of 300 miles providing much useful data on zero-g effects on living organisms.

Short-term experiments in weightlessness were also performed by human subjects in aircraft. To induce weightlessness the aircraft had to be precisely controlled to fly a powered parabolic path where engine thrust and lift just counterbalanced the pull of gravity. Often it was possible to keep men weightless in the cabin for more than half a minute. This became a standard technique for astronaut training.

It was some time, however, before NASA was sufficiently confident to risk one of its human 'guinea-pigs' to the uncertainties of ballistic flight.

The very first Redstone to carry an unmanned Mercury capsule, on 21 November 1960, fizzled on the pad, lifting a few inches as engine thrust rose and then subsided. Fortunately, it did not topple.

What happened was this. As the Redstone was heavier than usual for this test, it rose more slowly, thus causing one of the two ground umbilicals to disengage 20 milliseconds before the other – they should have dropped out together. The abort sensors in the Redstone detected the anomaly and shut down the engine. A circuit was set up at this point by the unusual events which allowed the capsule's automatic system to take over as was planned for normal flight operation. It received the MECO signal and jettisoned the escape tower (supposed to happen at 56 km. (35 miles) altitude). Then, sensing its low altitude, it went through the recovery operations, i.e. broke out the parachutes, started the flashing light beacon and radio homing devices.

Only one operation was omitted – capsule separation. At Cape Canaveral the event is still remembered as 'the day they launched the escape tower'.

As the Redstone had been slightly damaged by recontact with the launcher the capsule was fitted to another rocket. A month later it made a textbook flight down the Atlantic Missile Range parachuting down into the sea some 236 miles from the Cape. Maximum height of the trajectory was 135 miles.

In the test launchings that immediately preceded the first American manned spaceflight, three stand out as significant. One occurred on 31 January 1961, when Ham, a 137-lb chimpanzee, was lobbed in the Mercury MR-2 capsule a distance of 420 miles. The performance was greater than planned because of 'excessive booster thrust', the maximum altitude being 155 miles, some 40 miles higher than envisaged in the flight programme. In fact, the capsule was driven at a speed greater than 5,000 m.p.h., much faster than intended. Improper working of the engine system, due to a jammed thrust regulator, triggered the emergency escape rocket near the end of boosted flight. This forcibly separated the capsule, adding further to the already increased velocity and range.

Despite the fact that a pressure bulkhead was punctured on landing, the capsule was safely recovered with Ham inside. The chimpanzee had been in a miniature contoured couch inside a pressure chamber. Oxygen was introduced into the chamber, and the expired air passed through the spacecraft's environmental control system where impurities were removed before it was recirculated with added oxygen. Bio-medical data were sensed by electrodes attached to the animal, measurements being telemetered to the ground and simultaneously tape-recorded.

Mercury chimpanzees, trained at the Aeromedical Field Laboratory, Holloman Air Force Base, New Mexico, were expected to depress two handles in response to light signals. Failure to perform was 'punished' by slight electric shocks. This helped to determine if simple tasks could be carried out during launching, weightlessness, and re-entry; also to discern psychological and physiological effects on work-tasks performed under stress. The animal was photographed by a 16-mm. ciné-camera through a transparent window in the pressure chamber.

Next in the launch programme came an Atlas booster firing on 21 February 1961, to check maximum heating under the worst possible conditions of re-entry. The capsule reached a peak altitude of 108 miles, re-entering the atmosphere at a speed of about 13,000 m.p.h. It landed 1,425 miles downrange.

On 25 April 1961 – just thirteen days after Gagarin had made his celebrated circuit of the Earth – America prepared to orbit an unmanned Mercury capsule for the first time. Standing in for a live astronaut was an astronaut-simulator, 'a breathing, sweating, and talking robot'. This placed the same loads upon the environmental control system as a man would have done. It removed oxygen and added carbon dioxide and water vapour to the recirculatory oxygen, in effect duplicating the human circulatory and respiratory functions. Electrical heating elements added 200 W to simulate body heat. Two playback tape-recorders, each with pre-recorded voice messages of 45 min. duration, were also fitted to evaluate the communications system. The messages by-passed the normal microphone and were relayed to ground stations over the spacecraft's transmitters.

The launching, however, was disastrous. After a smooth lift-off the Atlas booster began to veer from course; after 40 sec. the range safety officer was forced to press the 'destruct' button. The radio signal triggered explosives in the rocket which blew it up.

Once again, this test showed the effectiveness of the emergency escape system. Before the booster was destroyed, sensing devices registered the booster's erratic behaviour, immediately commanding separation of the capsule and ignition of the escape rocket. The capsule separated cleanly and descended into the sea beneath its parachute. Had a man been aboard he would certainly have survived.

Despite this setback NASA pressed ahead with plans to conduct their first manned test – with a Redstone booster. Although this was to be limited to a ballistic 'lob' down the Atlantic Missile Range, similar to that made with the chimpanzee Ham, it contained all the drama of a fully fledged orbital mission.

Astronaut Alan B. Shepard spent more than 4 hr. in his Freedom 7 capsule on 5 May 1961, while clouds dispersed and 'technical holds' were dealt with by launch technicians. At last, just after 0934 hr. EST, came the moment of truth and Shepard was away to a perfect launch.

The sequence followed exactly as planned. The Redstone made a programmed ascent, pitched over at the correct angle to accelerate away over the Atlantic and cut-off at T-plus 42 sec. It was then inclined at 40° to the horizontal, with a speed of about 4,500 m.p.h. and altitude 196,000 ft. After the escape tower had jettisoned, the capsule was separated by releasing a clamp ring and firing three small rockets mounted in the retro-rocket pack attached to Mercury's heat shield.

Next the capsule's periscope was extended. Then, according to programme, the automatic attitude control system turned the craft round so that the blunt heat shield was pitched up at 14·5° to the path of flight.

It was at this point that Shepard first demonstrated man's ability to control a spacecraft under weightless conditions. By working a control handle he gently pitched, yawed, and rolled the craft to a maximum deviation of 20°, using bursts from gas-jet nozzles.

Shepard himself pitched up the capsule within 5° of the required angle of 34° prior to firing the three 1,160-lb.s.t. retro-rockets. When the spacecraft reached its peak altitude at 116 miles, these were fired in a 30-sec. burst, one after the other, reducing speed by some 350 m.p.h. Then the retro-rocket pack was jettisoned by detonating an explosive bolt securing steel straps.

After retracting the periscope, Shepard restored the capsule to automatic control – although he could still 'override' this with the hand controller. Finally, just before re-entering the atmosphere, he made a touch test with his right hand of certain equipment in the cabin. All the time he was under observation from a 16-mm. ciné-camera mounted above and behind his left shoulder.

Following re-entry – when Shepard experienced a maximum deceleration of above 5 g for about half a minute – the drogue parachute was ejected with radar chaff at 25,000 ft altitude. Thirty seconds later the main parachute deployed lowering the capsule into the sea at 32 ft/sec. Impact with the water was cushioned by a landing bag released from the capsule when the heat shield was detached during the landing sequence.

The records show that Shepard travelled a distance of 297 miles;* the flight lasted 15 min. 22 sec. of which he was weightless for 4 min. 45 sec.

* Earlier NASA gave the distance travelled as 302 miles and peak altitude 115 miles.

Further valuable experience in controlling a Mercury capsule was obtained by the late Captain Virgil 'Gus' Grissom on 21 July 1961, when he performed a similar sub-orbital flight in his Liberty Bell capsule. After separating from the Redstone booster the craft reached a height of 118 miles and splashed down 303 miles downrange. As in the case of his predecessor, Grissom manually controlled the capsule's 34° attitude for firing the retro-rockets. In this he was aided by a pilot-observation window which replaced two 6-in. portholes in Freedom 7. Reference lines inscribed on the four-pane window allowed the astronaut to align the craft precisely with the horizon, and the 'picture-window' also gave a much-improved view of the Earth and the stars. Measuring 19 in. across the top it was located directly above the astronaut on the capsule's centre line.

It was after Liberty Bell had come down beneath its parachute that disaster nearly struck. As Grissom waited inside the capsule to be picked up, suddenly the escape hatch blew out and the capsule immediately began to fill with water. Grissom struggled out and had to swim for several minutes before he was rescued by a helicopter. Meanwhile, the crew of another helicopter – a Marine Sikorsky HUS-1 – had managed to get a line attached to the sinking capsule, but the struggle to save it proved too great. At one point the HUS-1 had a wheel in the water, so determined were the crew to hold on to it until a rescue ship arrived; then a red warning light (false, as it later transpired) appeared in the cockpit indicating overheating of the helicopter's engine. Connection with Liberty Bell was severed and it sank in 18,000 ft of water.

Meanwhile, Grissom was taken aboard the USS *Randolph*. Although he had swallowed a lot of water, he quickly recovered. Later he explained that he had removed the safety pin from the escape hatch a short time before the explosive bolts which secured it blew out; apparently there had been a short-circuit.

Despite the success of the two manned sub-orbital shots, NASA had yet to get a Mercury spacecraft into orbit. On 13 September 1961, a capsule carrying an astronaut-simulator was boosted into space on the nose of an Atlas; after separating it was successfully orientated by ground control; the retro-rockets fired as planned near the end of the first orbit, and it landed in the sea about 160 miles east of Bermuda.

Then, on 29 November, came the turn of chimpanzee Enos to make an orbital flight in an Atlas-boosted spacecraft. The orbit achieved, inclined at 32°30′ to the Equator, ranged between an altitude of 100 and 147 miles. However, although intended to perform 3 revolutions of the Earth, telemetry signals indicated overheating and an orientation fault affecting roll control. The capsule therefore was commanded down after 2 orbits; it made a successful return into the Atlantic 220 miles south of Bermuda.

Now, at last, came the opportunity to launch an American into orbit. The man was Lieutenant-Colonel John Glenn of the US Marine Corps; his 3 orbits of the Earth on 20 February 1962 were a milestone in the development of manned spaceflight. However, the mission was not without its problems and it could well have ended disastrously.

After spending 3 hr. 44 min. in the Friendship 7 capsule on the launch pad, due to a series of 'technical holds', Glenn lifted off at 0947 hr. EST. The Atlas-D booster made its pre-programmed ascent over the Atlantic and at 1000 hr. came confirmation that the capsule had been successfully injected into orbit. Glenn was orbiting between 100 and 163 miles above the Earth, with an inclination of 32·5° to the Equator; this was declared good for at least 7 circuits although only 3 were attempted.

All went smoothly during the first revolution. Having entered orbit 503 miles downrange from Cape Canaveral, 20 min. later he was leaving the east coast of Africa to enter the night side of the Earth over the Indian Ocean. At 1025 hr. he consumed food paste from a tube, and at 1043 hr. he was in contact with the tracking station at Muchea, Australia, reporting he had sighted the 'bright lights' of Perth. At 1050 hr. Mercury Control gave the reassuring information that Glenn's heart-beat and respiration were 'completely normal'.

Just after 1100 hr. while approaching the Californian coast, he watched a glorious sunrise to be greeted with the strange sight of 'thousands of luminous particles' around the capsule.

Friendship 7 completed its first orbit at 1121 hr. at an average speed of 17,545 m.p.h. At 1128 hr. Glenn reported difficulties with the automatic attitude control system. The capsule was drifting off in yaw to the right at about 1°/sec. After yawing about 20°, it swung back to zero.

Glenn switched to the Fly-by-Wire system which allowed him to take overriding manual control. By manipulating the control handle he found he could stop the yawing motion and so conserve precious hydrogen peroxide fuel. After the flight it was suspected that the feed to a yaw attitude control nozzle had become clogged causing the capsule to yaw repeatedly to the right. When the 20° dispersion limit of the system was reached, the capsule was pushed back by the high-thrust jet.

The big problem, however, was yet to come. As Glenn was on his third and final orbit, the Muchea ground station picked up a telemetry signal indicating that the capsule's heat shield might have become detached due to a faulty switch. The reason for having a detachable heat shield was related to the final stages of recovery, when the capsule was suspended from its parachute just before coming down in the sea. The shield dropped down 4 ft pulling out a perforated skirt of rubberized glass-fibre; this formed an air-cushion to mitigate the landing impact.

If this shield became detached in orbit there would be nothing to prevent the craft from burning up as it re-entered the atmosphere. Mercury Control calmly discussed the position with Glenn. It was decided to keep the retro-rocket pack on the centre of the heat shield as the spacecraft re-entered the atmosphere. Normally, of course, the retro-rocket pack was jettisoned after retro-fire, being secured to the base of the capsule over the heat shield by three metal bands, released by detonating a single explosive bolt at its centre. But if the shield *had* become detached, these bands might serve to hold it in position long enough for air pressure to retain it.

Glenn described this frightening situation when I interviewed him for BBC's Ten O'Clock programme. After retro-firing at 1420 hr. EST, with the capsule some 600 miles west of Los Angeles, the first effects of re-entry became apparent some minutes later.

As the capsule began to feel the effects of frictional heating suddenly there was a 'bump'. Apparently this was a steel retaining strap breaking, but at the time Glenn thought the retro-pack had jettisoned. As heating increased an orange glow appeared outside the cabin, and it quickly became apparent that something was disintegrating on the outside of the shield. Flaming chunks of debris up to 8 in. across, glowing bright orange, streamed past the window.

As Glenn believed the retro-pack had gone, he wondered if the heat shield itself might be tearing up. In fact it *was* the retro-pack, and happily Glenn survived to tell the tale. His spacecraft splashed down in the Atlantic some 210 miles north-west of San Juan, Puerto Rico, at 1443 hr. EST having covered a total distance of 80,966 miles. Ironically, it was later discovered that the fault had been illusionary; the heat shield was securely attached all the time.

Examining this capsule after the flight one could judge for oneself the rigours of the mission. The heat shield was pitted and charred, flaking in layers, but still there was a good inch of ablative material round the edge. The outer skin of the capsule, though blackened by heat, was structurally sound. This was much the same as had been experienced on previous unmanned orbital missions. Normally, the shield was subjected to a maximum temperature of around 1,650°C at 25 miles altitude, with the capsule moving at nearly 15,000 m.p.h.; the conical section of the spacecraft behind the bow wave of plasma received a maximum of 927°C.

Glenn's capsule, much lighter than the Soviet Vostok, weighed 4,265 lb at lift-off. By the time it had arrived in orbit, mainly due to removal of the escape tower, this had reduced to 2,987 lb. At the time of recovery it was 2,422 lb. Two main factors accounted for the greater weight of the Soviet vehicle. It had a re-entry capsule of larger internal volume, and it employed more rugged constructional techniques. This was partly due to the Soviet decision to use standard atmosphere in the cabin at normal pressure, whereas American designers had adopted an all-oxygen system at 5·5 p.s.i. pressure (see page 264).

Despite Glenn's preoccupations with the capsule, he calmly reported instrument readings to ground stations, and made observations of the Earth and the stars. He remarked on different patterns in ocean currents, such as the Gulf Stream, and he described in some detail the area north-west of El Paso, Texas, remarking on the large area of desert and its squares of irrigated land.

The enforced tasks of this flight proved beyond doubt the ability of a well-trained astronaut to work under zero-g conditions as an inherent part of the spacecraft system. Glenn himself thought it might be possible to use considerably less automation on future flights and this was amply demonstrated in the later Gemini programme when the first human-controlled spacecraft were manœuvred in missions involving orbital rendezvous.

Three months later another Mercury capsule stood ready on its Atlas booster at the Cape. This was Aurora 7 with Lieutenant-Commander M. Scott Carpenter. Launched at 0845 hr. EDT on 24 May 1962, the craft orbited between altitudes of 100 and 169 miles at 32·5° inclination. Carpenter had been instructed to use the Fly-by-Wire system, in which he had overriding manual control, and to keep the purely automatic system in reserve.

The first problem encountered concerned overheating of the pressure suit. After exercising with a rubber cord the suit temperature reached 27·8° C during the first orbit. Carpenter was asked to relax his efforts, to get the temperature down; he did so and it gradually dropped below 21° C.

While still on the first orbit over the Western Pacific, the astronaut switched off the orientation system to conserve fuel, allowing the spacecraft to drift in the manner of an unstabilized satellite. This left him free to concentrate on other tasks. After taking his first paste meal from a tube, he looked for Glenn's mysterious luminous particles. As the capsule approached the US coast from over the Pacific, he noticed some, like 'snowflakes', but they were not in the numbers Glenn had described.

Apart from instrument readings and a heavy programme of photography, which included separation of the Atlas stage, features of the Earth's surface and cloud cover, and luminous particles, Carpenter had two important experiments to perform, turning his craft into a rudimentary physics laboratory.

The most elaborate involved a Mylar plastic aluminized balloon 20 in. in diameter, which had to be released from the capsule at the beginning of the second orbit, on the end of a 100-ft line. A mechanical arm, to which the line was anchored, was designed to measure any slight pull on the balloon due to drag effects. The balloon – which had five panels of different colours – was meant to help estimate distances in space and also to determine the most easily recognizable colour under different conditions of illumination. This, of course, had much to do with forthcoming experiments in orbital rendezvous and docking.

However, although the balloon deployed as planned, it did not fully inflate. Carpenter reported no effect on the spacecraft's drift; the line was sometimes taut and sometimes slack. Most brilliant when illuminated by the Sun were the orange and silver segments. The other panels on the balloon were white, yellow, and phosphorescent, the last being used to assist observation at night. A number of multi-coloured $\frac{1}{4}$-in. diameter Mylar discs, released from the folds of the balloon, were used to provide a comparison with the luminous particles. The balloon, having failed to release in orbit, trailed behind the spacecraft as it re-entered and burned up.

As a further aid to the Gemini programme, an experiment was also provided for the observation of liquid both in a weightless environment and when the retro-rockets were fired prior to re-entry. Mounted in the cabin was a 3-in. diameter flask with a central stand-pipe; three holes in the base of the pipe allowed passage of the liquid which comprised distilled water, green dye, an aerosol solution to reduce surface tension, and a silicone additive to inhibit foaming.

The flask was filmed by the astronaut-observer camera. Under zero-g, the liquid was expected to rise in the stand-pipe due to surface tension, rather than float about in globules. The results were related to the design of spacecraft tanks and restartable engine systems.

Thus, in many ways, the flight of Aurora 7 was a demonstration in miniature of the function of a manned orbiting laboratory.

However, Carpenter had his share of troubles. Apart from overheating of the spacesuit during the first and second orbits, a fault developed in the pitch horizon scanner and inadvertent use was made of two attitude control systems simultaneously, including the high-thrust system. Fuel was conserved by allowing the spacecraft to drift for 77 min. beyond the time scheduled in the flight plan.

Difficulties with the automatic attitude system meant that Carpenter had to pitch up the craft at the correct angle for retro-fire using the manual controls. Radar tracking data indicated that, in the period approaching retro-fire, the spacecraft 'had an average yaw error of 27°'. This, combined with late ignition of the retro-rockets, resulted in the capsule overshooting its scheduled landing point in the Pacific by 250 nautical miles. Carpenter pushed the manual retro-fire button when the retro-rockets did not fire automatically. This was 3–4 sec. later than planned, which accounted for 15 to 20 miles of the total overshoot error.

Then, despite the fact that a new radio frequency was being used in an attempt to defeat the radio blackout caused by the ionized sheath of plasma which forms around a spacecraft as it re-enters the atmosphere, radio contact was lost. While the world anxiously awaited news of the astronaut, at last came word that Carpenter had been located some 125 miles north-east of Puerto Rico. Having landed at 1741 hr. EDT, he had emerged from his capsule and awaited pick-up in his emergency dinghy. The flight, lasting 3 orbits, covered a total distance of 81,325 miles.

Lessons learned by Carpenter were quickly applied to the next Mercury mission. They included provision of a switch to isolate the high-thrust attitude control jets and revised fuel management training procedures.

Next to go into space from Cape Canaveral was Commander Walter M. Schirra. Lifting off at 0815 hr. EDT on 3 October 1962, he travelled in the capsule Sigma 7. His initial orbit ranged between 101 and 177 miles.

The mission, planned to last nearly 6 orbits, allowed more time for observations and experiments. A primary aim was to discover how much fuel in the H_2O_2 orientation system could be conserved. As Schirra started on his third orbit he reported that the capsule still had 90 per cent of its fuel supply. The capsule was allowed to drift in attitude with all controls switched off for 99 min. of the flight programme. The only anomaly was elevated suit temperature experienced during the first 2 hr.; this was later attributed to a foreign substance in a control valve.

Of the experiments, three high-intensity flares ignited at the Woomera rocket range were not directly observed because of intervening cloud cover. It had been hoped that, with the aid of a photometer, Schirra could establish the degree of atmospheric attenuation of a light source of known intensity. All Schirra saw above the clouds was a diffused 'block of light'.

Many photographs of the Earth were obtained using a 35-mm. ciné-camera. Scientists, having examined pictures taken by previous astronauts, were interested in obtaining colour photographs of fold mountains, fault regions, volcanic fields, meteorite impacts, and glaciers. They were also keen to study the photometric properties of various land surfaces for comparison with features of the Moon and other planets.

Schirra also saw fluorescent particles, or 'snowflakes'; he said he could produce them by banging on the walls of the cabin. Another subject of interest was the type and magnitude of the interaction of nuclear particles in the space environment; two radiation-sensitive emulsion packs were mounted on each side of the astronaut's couch.

Test samples of heat-resistant materials were bonded to the cylindrical neck of the capsule for post-flight examination. Some had discrete cracks or slots, half of which had been filled or repaired. In this way it was hoped to discover the effectiveness of possible repairs made to spacecraft structures.

Apart from a 2-sec. delay in retro-fire, the spacecraft made an uneventful return into the Earth's atmosphere, landing in the Pacific Ocean some 295 miles north-east of Midway Island at 1728 hr. EDT. The distance covered was about 153,900 miles.

Last in the Mercury series was the most ambitious and in many ways the most successful. It was planned as America's first day-long spaceflight, with the aim of achieving 22 orbits. With astronaut Major L. Gordon Cooper at the controls of capsule Faith 7, lift-off came at 0904 hr. EDT on 15 May 1963, with the vehicle entering an orbit ranging from 100 to 166 miles.

As in the case of astronauts Glenn, Carpenter, and Schirra, Cooper had many control tasks to perform. Also, because of the prolonged period spent under weightlessness, special importance was attached to medical studies.

The first orbit was largely devoted to checking proper function of the spacecraft and radio communication with the ground. Cooper observed the lights of Perth, Australia, and received the 'go' signal for 7 orbits. However, irregular behaviour of the pressure suit heat exchanger on this and subsequent orbits led to abandonment of the prearranged 8-hr. rest programme. Cooper said he dozed periodically for 10 to 15 min. throughout the mission.

On the second orbit blood-pressure was measured and the astronaut performed exercises using elastic cords. He took his first 10-min. nap. On the third orbit a battery-powered 5·75-in. diameter sphere, with two flashing xenon lights, was released from the retro-pack into space. The lights flashed at about once per sec. and Cooper was able to observe them on the night side of the fourth and fifth orbits. Again this experiment, related to later Gemini and Apollo missions, helped to judge distances in space. As the light intensity and separation speed were known, it was possible to obtain visual sighting data up to about 15 miles.

At the beginning of the fifth orbit, a cabin temperature test was made with the coolant system inoperative, while the astronaut relied on the suit cooling circuit alone. This provided engineering data for the design of future heat exchanger systems.

Near Bloemfontein in South Africa a 3,000,000-candlepower xenon light, switched on for 3 min., was observed on the fifth orbit; also lights of the near-by city. This was mainly to check the feasibility of using ground or high-altitude lights as navigation fixes for mid-course and near-Earth corrections in project Apollo;

also to give some indication of light attenuation through the atmosphere. On a number of orbits radiation measurements were taken; and during the sixth the balloon experiment, only partly successful in the Carpenter mission, was tried again. This time the multi-coloured Mylar balloon failed to eject.

Meals on the extended Mercury mission were more varied. There was 'ready-to-eat', bite-sized food 'in sufficient quantity to satisfy all calorific requirements', and the experimental, Gemini-type, dehydrated food and drink prepacked in plastic containers for reconstitution during the flight.

Preparation of the dehydrated food required the addition of water. The containers were fitted with nozzles through which water was added; food (or drink) was forced out of the same nozzles after hydration. The drink was ready for consumption shortly after water was added; the food required about 5 min. of mixing. Cooper carried a food supply totalling 2,376 calories, including such meals as spaghetti and meat sauce and a beef and gravy dinner.

After taking refreshment during the seventh orbit and exercising, he received the 'go' signal for a further 10 orbits. Between the tenth and thirteenth orbits Cooper slept, awakening on the fourteenth without a signal. After making routine reports he ate and drank, and an orbit later began a series of observations including dim-light phenomena and horizon definition photography. Cooper was then given clearance for the scheduled 22 orbits.

More observations were made as the flight progressed, including infra-red weather photography. Cooper also saw the luminous 'snowflakes' but concluded they were created by moisture particles formed by hydrogen peroxide vapour from the spacecraft's attitude control nozzles.

But by far the most remarkable observations reported by Cooper were those concerned with detail on the ground which, according to the laws of optics, were 'impossible'. Apart from seeing rivers, lakes, mountains, and islands while over North Africa, he saw what he thought was the wake of a boat on the Nile. Looking down from his capsule over Northern India and

Tibet, he spotted winding roads, wisps of smoke from chimneys, houses, road vehicles, and a train.

Space officials were incredulous and after the flight there was talk of space-illusions and hallucination. It was not until Cooper's subsequent Gemini mission with 'Pete' Conrad that the arguments were finally settled.

Meanwhile, coming near the end of his prolonged orbital flight, Cooper had other matters to concern him. Difficulties with the spacecraft's orientation system necessitated him taking over manual control. The trouble arose with two connections to an amplifier-calibrator which served to convert electrical signals of various spacecraft systems into commands. Part of this unit's function was to relay commands to activate H_2O_2 thrusters in the automatic attitude control system. The fault was subsequently traced to moisture in the spacecraft causing corrosion in and around electrical connections.

Cooper pitched up the spacecraft using the manual controls and fired the retro-rockets. After the electrical failure, he was also instructed by Mercury Control to complete the re-entry phase on manual. Instructions for retro-sequence checks, manœuvres, and timing were provided by astronaut John Glenn acting as spacecraft communicator aboard the tracking ship *Coastal Sentry* offshore of Japan.

Splash-down came about 80 miles south-east of Midway Island in the Pacific at 1924 hr. EDT on 16 May. The capsule was clearly seen descending by parachute from the aircraft carrier *Kearsarge* which secured recovery some 37 min. later. Cooper had travelled a distance of 583,469 miles.

Despite significant progress made in the Mercury programme, America was trailing badly in man-hours spent in space. Of the six US astronauts only four had made orbital flights lasting a total of 53 hr. Between them they had covered a distance of 900,000 miles. In contrast Soviet Vostok cosmonauts had logged 382 hr. Valentina Tereshkova alone had been in space 17 hr. longer than all the US astronauts put together.

It was some two years before the Americans began to reverse this situation in the brilliantly executed Gemini programme.

Once again McDonnell Aircraft Corporation was made prime contractor by NASA. This time the vehicle had to sustain two men, with the astronauts having full manual control of orbital manœuvre. Moreover, by offsetting the vehicle's centre of mass, it was required to generate a proportion of aerodynamic lift during re-entry giving the ability to steer the vehicle more precisely towards a fixed landing point. Gemini was named after the constellation containing the two stars Castor and Pollux, the 'heavenly twins'.

After extensive ground testing the launch programme began on 8 April 1964, when a two-stage Titan 2, adapted from the ICBM – which now took over manned spacecraft launching from Atlas – sent an unmanned Gemini into orbit from Cape Canaveral.

This initial test was purely to evaluate launch-vehicle/spacecraft compatibility. There was no attempt to separate the spacecraft from the rocket's final stage. The orbit, angled at 33° to the Equator, was slightly higher than expected; it ranged between 100 and 204 miles. Nevertheless, all flight objectives were met and 104 instrument measurements were received during the first orbit.

The next flight, still unmanned, came on 19 January 1965, when a Gemini capsule was launched over a sub-orbital trajectory, this time separating from the rocket in order to test the capsule's re-entry behaviour. After reaching an altitude of some 99 miles and surviving the rigours of frictional heating the craft parachuted down into the Atlantic and was recovered.

So successful were the results of these preliminary tests that the first manned Gemini flight (GT-3) was launched on 23 March 1965. Lifting-off from Cape Canaveral at 0924 hr. EST, following 'technical holds' lasting only 24 min., the initial orbit, angled at 33° to the Equator, ranged between 100 and 139 miles. Aboard were the late Virgil I. Grissom, command pilot, and John W. Young. Orbiting the Earth three times in 4 hr. 53 min., theirs was the first manned spacecraft to be manœuvred out of one orbit into another. This was achieved by using the high-thrust control rockets aboard the spacecraft worked by manual controls.

How this affected succeeding orbits can be seen from the fact that on the second orbit, perigee and apogee distances were 98–105 miles, and on the third 100–140 miles. Adjustment of orbital height or small plane changes were accomplished by controlled bursts from 85-lb and 100-lb thrust chambers working in conjunction with 25-lb thrusters providing control in pitch, yaw, and roll. The propellant for all these motors was nitrogen tetroxide (oxidizer) and monomethylhydrazine (fuel).

Splash-down came at 1417 hr. EST in the Atlantic Ocean not far from Grand Turk Island. The distance travelled was about 80,000 miles. In this mission Grissom, who had previously made the second sub-orbital Mercury flight in 1961, became the first man to enter space for a second time.

Next into orbit were James A. McDivitt and the late Edward H. White II. Lift-off occurred at 1016 hr. EST on 3 June 1965, with the Gemini 4 spacecraft entering an orbit ranging between 100 and 175 miles. One of the first objects of this mission was to rendezvous with the second stage of the Titan launch vehicle, after separation of the spacecraft, using Gemini's ability to manœuvre under rocket thrust. This proved abortive. McDivitt said the rocket casing fell away too rapidly; it was also turning over at 40 to 50 deg./sec. – 'much faster than anyone had anticipated'. In a vain attempt to chase the booster about 42 per cent of the available fuel was consumed. It was suggested that a successful rendezvous might depend on the use of radar and a larger fuel supply. When the manœuvres were finally called off, the spacecraft was orbiting between 103 and 182 miles above the Earth.

The planned extra-vehicular activity of astronaut White was delayed from the second to the third orbit to allow more time to prepare for the experiment. In the pure oxygen environment of Gemini's cabin at 5 p.s.i., McDivitt helped White put on his special equipment, which included an emergency oxygen chest pack. Finally, both astronauts pressurized their suits at 3·7 p.s.i.

After the cabin had been depressurized, at 1442 hr. EST the hatch was opened on White's side. Three minutes later the astronaut climbed out of the hatch and using a hand-held gas-gun gently launched himself into space, his only restraint being a 25-ft umbilical tether. The tether also contained the oxygen supply hose and electrical leads; it had a Mylar aluminized and nylon covering.

White found manœuvring with the gas-gun comparatively simple, but as its fuel was exhausted after 3 min. he spent the remainder of his time 'getting the feel of space' by twisting his body, moving his arms and legs, and pulling on the tether to induce tumbling motions. There were no disturbing reactions and, in fact, White having described the experience as 'exhilarating' remained outside longer than planned – 21 min. His pulse at the start of EVA was 150 and just before re-entering the spacecraft it was 178. After some difficulty in resealing the hatch, it was finally closed at 1506 hr. and the cabin repressurized.

The flight of Gemini 4 was also significant in other ways. On the Earth White reported seeing roads, boat wakes, strings of street lights, airfield runways, and smoke from trains and buildings. Many observations and tests were made during the 4-day mission, not least of these being related to bio-medical effects. At the time, particularly with regard to Soviet results, there had been some anxiety concerning prolonged exposure to weightlessness and its influence on the human cardiovascular system. However, although both men had temporarily reduced blood-pressure after the flight, medical tests gave no indication that adequately prepared people could not safely endure spaceflights of longer duration.

Although a fault in the spacecraft's computer did not allow the spacecraft to make a semi-lifting re-entry, the ballistic return was sufficiently accurate to place the vehicle some 45 miles from the appointed landing place 390 miles east of Cape Canaveral and 230 miles north of San Salvador. The splash-down came at 1212 hr. EST on 7 June, the spacecraft having made 62 revolutions* of the Earth and covered a distance of 1,609,700 miles.

The next mission, planned to last 8 days, was to investigate the problems of orbital rendezvous in more detail. It was also the

* At this time NASA began to count Earth revolutions rather than orbits, this being defined as the number of passes over the longitude of the launch site.

first mission to use fuel-cells instead of chemical batteries to supply electrical power. Aboard Gemini 5 were L. Gordon Cooper, Jr., and Charles Peter Conrad. This time there were no 'technical holds' and the Titan rocket lifted the capsule off the pad at precisely 0900 hr. EST on 21 August 1965, achieving an initial orbit of 101 to 217 miles.

First task was to release from the spacecraft's adapter module a 76-lb Radar Evaluation Pod (REP). This device embodied the same type of radar system then being developed for incorporation in Agena spacecraft for full-scale rendezvous and docking experiments. Although the target was released, rendezvous manœuvres were frustrated by troubles which developed in Gemini 5 affecting the fuel-cell electrical power supply. The fuel-cells themselves, developed by General Electric, worked perfectly; the trouble stemmed from a simple heater fault in the cryogenic oxygen tank. This led to excessive heat transfer into the hydrogen tank, and it was feared that the cells would produce too much water (the by-product of the reaction) which could swamp the system. After the trouble first arose, probably during the launch period, tank pressure had dropped from 850 to 60 p.s.i.

Nevertheless, the Radar Evaluation Pod was released shortly after entering the night side of the Earth on the second orbit and a number of infra-red and radar measurements were made. However, the pressure drop in the spacecraft's fuel-cells became so alarming that the planned manœuvre exercises had to be abandoned. So serious was the situation that emergency plans were made to return the spacecraft during the sixth orbit. Meanwhile, NASA and McDonnell teams estimated that if the oxygen tank pressure could be stabilized at around 60 p.s.i. it would gradually build up from that point. McDonnell engineers actually demonstrated the restorative powers of a duplicate set of fuel-cells under simulated space conditions as the flight progressed, and the astronauts were given the 'go' signal for another day in space on the strength of the results. As predicted, oxygen tank pressure built up steadily as the mission progressed, and the scheduled 8-day flight was allowed to proceed to its close.

The radar target by now had been lost in space. However, on

the third day, ground control arranged a rendezvous exercise involving a 'phantom' target. The astronauts demonstrated they could arrive at a given point in space with an accuracy of 0·2 miles within 2 min. of the planned arrival time.

But more troubles were in store. Excessive heat transfer into the hydrogen reactant tank, causing it to vent for the latter part of the mission, led to the spacecraft building up yaw and roll rates which at times were as great as 12 deg./sec. These had to be corrected by the astronauts using the attitude control jets.

On the sixth day of the flight two left yaw thrusters became inoperative probably due to freezing of the nitrogen tetroxide supply. Later, four more attitude control thrusters stopped working, it is believed due to fuel depletion.

But troubles dogged the mission right to the end. Ground errors, fed into the spacecraft's computer, affecting the craft's inertial co-ordinates at the time of retro-fire, led to the spacecraft falling short of its planned landing point by 103 miles and some 10 miles to the right of the ground track. However, the craft made a satisfactory approach, splashing down at 0756 hr. EST on 29 August some 335 miles south-west of Bermuda and about 760 miles east of Cape Canaveral.

Gemini 5 completed 120 revolutions of the Earth during this epic mission, at last surpassing the 119-hr. spaceflight performed by the Soviet cosmonaut Valery Bykovsky in Vostok 5. The total distance travelled was 3,338,000 miles.

An enormous amount of information was put on file by the Gemini 5 astronauts despite all their frustrations. Not least was final and overwhelming proof of the ability to pick out small objects on the Earth's surface. At the Sixteenth Congress of the International Astronautical Federation, held in Athens a month later, I asked Cooper about the smallest detail he had been able to discern from orbit, reminding him of the scepticism which greeted his previous observations from the Mercury spacecraft. He said it was possible even to see the wakes of ships at sea. One of the vessels identified was a recovery ship in the Atlantic. Conrad chipped in: 'This time he had a witness.'

At the Space Congress Conrad showed photographs, taken with specially adapted commercial cameras, which also contained remarkable detail. A score of beautiful colour shots illustrated areas of the United States, Mexico, Cuba, China, Tibet, Greece, and Crete. Taken through the cabin window with the nose of Gemini tilted straight down they were obtained with a hand-held 70-mm. Hasselblad. Surface photographs of still higher resolution were made with a 35-mm. Zeiss Contarex fitted with a Questar telephoto lens with a focal length of 56 in.; a system of mirrors 'folds' the light beam into a barrel 8 in. long. In this case the camera was fixed to a bracket behind the spacecraft's right-hand window.

The astronauts had been given a list of subjects to photograph, all within the United States and Africa. They included selected cities, railways, roads, harbours, rivers, lakes, illuminated sites on the Earth's night side, ships, and wakes.

High-resolution pictures taken over Cape Canaveral showed launch sites and earthworks connected with project Apollo in surprising detail. They were detailed enough to show a causeway and a bridge across the Banana River linking Cape Canaveral with Merritt Island. Conrad said they were able to pick out every launch pad along the coast with the naked eye, including 'good old pad 19' from which their own mission had begun. At one point a white line was identified as the freeway into El Paso. If you use a magnifying glass on the picture you can see Biggs Air Force Base and El Paso International Airport. 'Although we were 100 miles up,' Conrad said, 'each morning we could see the same aircraft of a scheduled flight going into an airport by picking up its contrail. One day we even saw an aircraft ahead of its contrail.'

But even more remarkable was the astronauts' ability to spot two Minuteman ballistic missiles deliberately launched on test as Gemini passed near US launch centres. Cooper said they saw smoke and flame from one missile as it pierced the cloud deck. For a time they lost the trail and then picked it up again in the airglow and watched it to the point of burnout. The second Minuteman, seen at a slant-range of 250–300 miles, was also located by its engine flame. After losing it for a moment they found that the Sun threw a shadow of the contrail against the cloud tops.

Orbiting above Holloman Air Force Base they spotted the plume of spray thrown up by a rocket sled during water braking. The sled, too, had been fired to synchronize with their orbital pass.

Whereas previous estimates for the resolving power of the human eye, with black and white contrast, were about 1 min. of arc, these observations demonstrated an ability to resolve a half a minute or less. Before men went into space a minor degradation in visual acuity had been forecast. Similarly, photographs taken from orbit, rather than being less definitive, have tended to be more distinct than pictures taken from high-flying aircraft. Indeed, geological features in the Earth's crust show up clearly in pictures obtained from orbit which would go unnoticed in photographs taken from the air.

On the basis of Cooper's original Mercury observations his visual acuity from orbit had been measured as 20/12 on the Snellen scale, conventional 'perfect' eyesight on Earth being 20/20.

Several explanations have been offered concerning the ability to perceive objects subtending such minute visual angles. According to Dr Eugene B. Konecci of the National Aeronautics and Space Administration, one concerns the possibility that the Earth's atmosphere acts as a huge refracting medium leading to vastly improved visual observation of objects on Earth directly below the orbiting astronaut. A second theory indicates that objects with long extensions (such as aircraft vapour trails, railway tracks, and smoke rising from chimneys) facilitate observation of tiny objects at the end point. A third explanation is involved with an integration of stimuli in the central nervous system. Possibly all three theories may be jointly involved.

After the 8-day mission of Gemini 5, at last came the day when a full-scale rendezvous would be attempted with an Agena target vehicle. Some 90 min. after the Atlas-Agena had been launched on 25 October 1965, Gemini 6 was to follow with the aim of making contact on the third orbit.

As astronauts Walter M. Schirra and Thomas P. Stafford waited in their spacecraft following the launch of the target rocket, downrange tracking stations reported loss of contact. Apparently, the Agena stage, having separated from the Atlas booster, had failed to inject itself into orbit.

The launching of Gemini 6 was promptly abandoned; and as another Agena target vehicle was not available, plans were made for two manned Gemini spacecraft to perform a double flight; this to include rendezvous.

Gemini 7, to be launched first with astronauts Frank Borman and James Lovell, was to attempt a 14-day mission as a final test of man's long-term adaptability to the space environment.

Lift-off came at 1230.03 hr. EST on 4 December 1965, again without 'technical holds'. The initial orbit ranged between 100 and 204 miles angled at 28·9° to the Equator.

Apart from extensive bio-medical tests the crew were to conduct some 20 experiments, 14 of which were repeats of those conducted on former missions. For the first time new lightweight removable pressure suits were worn; each weighed only about 16 lb. Three of the medical tests included study of calcium balance, two were navigation tests in preparation for project Apollo, and there was also a laser experiment. The laser test failed as the astronauts could not identify the ground-based argon gas laser beacon.

In preparation for the rendezvous experiment, a rendezvous transponder (similar to that used in the REP and Agena target vehicle) was fitted in the small end of the spacecraft. Weighing less than 50 lb, this was designed to receive signals from the rendezvous radar system in Gemini 6 and return them at specific frequencies and pulse width.

After a heavy experimental programme lasting 5 days, Borman and Lovell made preparations for the arrival of Gemini 6 which, on the morning of 12 December, stood ready for lift-off at Cape Canaveral. They waited in vain; premature release of an inhibitor plug from the launch vehicle stopped the launching, and the scheduled 'meeting in space' was deferred for 3 whole days.

However, when at last lift-off came at 0837.26 hr. EST on 15

December, there were no pre-launch difficulties. The craft went unerringly into an orbit ranging between 100 and 162 miles, at the same orbital inclination as Gemini 7. The orbit of the first spacecraft had already been circularized at 185 miles altitude in readiness for the rendezvous experiment.

At the point of starting rendezvous manœuvres Gemini 6 was some 1,200 miles behind Gemini 7. Working from the basis of computed ground-tracking data astronaut Schirra began a series of thrust corrections using the manœuvre rockets some 94 min. into the flight. After 6 major impulses giving 14·2 to 60·8 ft/sec. velocity changes and several smaller corrective bursts of thrust, the two craft were in radar contact at a distance of 270 miles. From that point further bursts of thrust from the manœuvre motors brought Gemini 6 to within 120 ft of Gemini 7, 5 hr. 47 min. into the flight. Despite all the anticipated problems, the astronauts used only 175 lb of their fuel supply – about half the amount allocated for rendezvous manœuvres.

The two craft remained in close formation for 20·4 hr., during which time the astronauts could see and signal to each other through their windows. Although the craft had no provision for docking, the distance between them was repeatedly brought down to a few feet. At one stage they were within a foot of each other, but the astronauts had agreed before the flight that the spacecraft would not actually touch.

After the rendezvous experiment, the Gemini 6 astronauts modified their orbit in preparation for re-entry. It had been planned in advance that their mission would not last longer than a day, and the craft was fitted with conventional batteries instead of oxygen-hydrogen fuel-cells. Gemini 6 splashed down in the Atlantic at 1029.09 hr. EST on 16 December having completed 17 revolutions of the Earth in a flight extending 449,800 miles.

Gemini 7, meanwhile, still functioning normally, maintained its experimental programme. NASA later confirmed that 75 per cent of the prescribed tests and observations had been conducted. These included spotting from orbit – visually and with the help of instruments – a submarine-launched Polaris A-3, and the re-entry vehicle of a Minuteman 2. The radiation signatures were measured by a radiometer.

Although minor problems occurred affecting the fuel-cells and attitude control system, these did not hamper operations. The new spacesuits worked particularly well, and long periods were spent with them off. The astronauts agreed this added greatly to their comfort and mobility; and Lovell, who spent more time in this 'shirt-sleeve environment', not only slept better than his companion but had a lower average heart-rate.

Splash-down for the Gemini 7 astronauts came at 0905.06 hr. EST on 18 December. Their record-breaking mission of 206 revolutions had lasted 330 hr. 35 min. 17 sec., an enormous leap in human space endeavour. The landing 700 miles south-west of Bermuda, only 7 miles from the appointed target, was the most accurate in the Gemini programme so far. The total flight distance was an impressive 5,716,900 miles.

The next vital step in the Gemini programme was to achieve rendezvous and docking, the objective that had been frustrated the previous October by the loss of the Agena target vehicle. Gemini astronauts had been learning to master the technique with a Docking Trainer at the Manned Spacecraft Center at Houston, Texas.

On 16 March 1966, another Agena target was launched by an Atlas-D from Cape Canaveral, and this time headed into space to achieve the planned 185-mile circular orbit angled at 28·9° to the Equator. To ensure that both vehicles orbited in the same plane, the launch azimuth of Gemini-Titan 8 was matched with Agena's orbit. As the target vehicle arrived in the precalculated position in space 101 min. after lift-off, Gemini 8 with astronauts Neil A. Armstrong and David R. Scott was launched in pursuit. The time was 1142.02 hr. EST, and the objective a 3-day mission including another experiment – by astronaut Scott – in extra-vehicular activity (EVA).

Initially the flight went remarkably well. The achieved orbit ranged between 100 and 168 miles, and this was successfully modified to bring the two craft into close proximity on the fourth orbit. After a short period of station-keeping during the next orbit, Armstrong prepared to complete the docking manœuvre. The object was to nudge the small end of Gemini, by means of steering jets, into the docking collar of Agena when an index bar on the spacecraft should engage a 'V' notch or slot. In line with the slot on the Agena vehicle was the vertical antenna for the L-band radar used to 'home' the spacecraft on the Agena, and to control the Agena from the spacecraft prior to docking. The target vehicle had its own propulsion system and could be manœuvred either from the spacecraft or the ground.

If the small end of Gemini had entered the Agena docking collar successfully, clamps inside the collar should grab the Gemini cone and pull it into a latched position. If the match had not started perfectly, the design was such that both craft should turn slightly and conform. If the Agena was bumped away the astronauts would merely set up for another attempt.

Six hours and thirty-four minutes into the mission the world's first docking between two spacecraft was safely accomplished. Not only was this a physical attachment between two vehicles, of the kind anticipated between spacecraft in the Apollo Moon programme, but in the same action control systems were automatically coupled. Thus the Gemini astronauts could start up Agena's 16,000-lb thrust engine and perform manœuvres as a combined spacecraft. The engine could then be used to make extensive orbital manœuvres.

It was some 20 min. after the vehicles had docked that troubles arose in dramatic fashion. Armstrong had fed commands to the Agena and successfully accomplished a yaw manœuvre when the combined vehicle began to develop a fast rate of yaw and tumble. As it turned out this was caused by the uncommanded firing of one of Gemini's 25-lb roll thrusters. Armstrong separated from the Agena and after some difficulty turned the thruster off; control was regained by applying bursts of thrust from the re-entry attitude control system. When the spacecraft was finally brought under control it was found that nearly 75 per cent of the re-entry system's fuel supply had been used up, and ground control sent instructions to terminate the mission.

The emergency re-entry resulted in a landing on 16 March at

2223.08 hr. EST 691 miles south-east of Okinawa. This was within 3 miles of the planned landing position. Armstrong and Scott emerged from the spacecraft about an hour after splash-down and awaited rescue in their inflatable raft. They were picked up by the destroyer USS *Mason* at 0138 hr. next morning.

Instead of the scheduled 3 days the flight had lasted 10 hr. 42 min. 6 sec., covering a distance of 181,450 miles and completing 7 revolutions of the Earth.

McDonnell traced the fault in Gemini 8 to a short-circuit in the electrical system affecting operation of the solenoid which opened the fuel and oxidizer valve in the thruster.

Owing to the curtailed flight programme only a small amount of data was received from 3 experiments, and Scott's 'space-walk' could not take place. Apart from using a hand-held gas-gun to control his movements, he was expected to experiment with a 'torqueless tool', loosening and tightening bolts on a work panel on the outside of the spacecraft's adapter module. The experiment was related to the problems of repair and assembly of large space structures in orbit. This important evaluation had to wait for another occasion.

Meanwhile, ground control repositioned the Agena target vehicle into a higher orbit so that it could be visited on a later mission. It carried a small micrometeoroid impact detector which Scott had intended to retrieve.

The last four Gemini missions were designed to develop routine rendezvous and docking manœuvres, to demonstrate the ability of astronauts to perform simple work-tasks outside their craft, and generally to develop techniques ready for the subsequent three-man Apollo spacecraft. The first development versions of the Apollo command and service modules were already being launched from Cape Canaveral, unmanned, by the Uprated Saturn 1.

With Gemini 9 it was planned to make three rendezvous and docking tests with an Agena target vehicle and also to perform an EVA experiment lasting a total of 2 hr. 25 min. The latter was mainly to test an Astronaut Manœuvring Unit (AMU) fixed to the astronaut's back.

The first disappointment came on 17 May 1966, when an Atlas carrying an Agena target vehicle developed a control system fault and fell into the Atlantic. Once again a Gemini launching had to be postponed until another target vehicle could be prepared; and this time a 1,700-lb Augmented Target Docking Adapter (ATDA) stood in for the Agena.

The ATDA, specially developed by McDonnell for such an emergency, had a similar docking collar as Agena but did not embody a rocket engine for powered manœuvres. Nor did it have the micrometeoroid impact package as mounted on the Agena.

All went well with the launching of the ATDA on 1 June 1966, until telemetry reported that the nose-fairing had failed to eject. Nevertheless, it was orbiting successfully 186 miles above the Earth at the prescribed 28·9° inclination.

Then came a second frustration. An attempt to launch Gemini 9 some 100 min. later had to be abandoned because of a fault in ground data transmission affecting the rendezvous operation.

Astronauts Thomas P. Stafford and Eugene A. Cernan had to wait another 48 hr. for their mission to begin. This time, however, there were no 'technical holds' and on 3 June the rocket made a smooth departure at 0839.33 hr. EST to reach an orbit which ranged between 99 and 168 miles.

Orbiting in the same plane as the ATDA, there were no difficulties in achieving rendezvous which was completed during the third orbit. The astronauts then could appreciate the problem with the nose-fairing with their own eyes; it was hanging open 'like the jaws of an alligator' and shrouding the docking collar which Gemini was meant to engage.

Although Cernan was due to perform his first EVA experiment there was no suggestion that he should transfer to the vehicle and attempt to release the fairing.

Close inspection clearly showed that explosive bolts holding the fairing clamp ring had fired. However, the fairing was actually held not by the clamp but by wiring to the bolt firing squibs. The astronauts were able to make a full inspection from a distance of a few feet, reporting their findings to the ground. It appears that lanyards meant to disengage wiring quick-release plugs had not been properly attached prior to launching.

The flight plan was therefore revised to demonstrate two re-rendezvous manœuvres during the first day without the use of radar. These tests also simulated emergency docking manœuvres in which an Apollo Lunar Module (in this case Gemini) had to dock from a position above the target vehicle.

It was on the second day – on the thirty-first revolution – that Cernan began his EVA mission. His period of stay outside the spacecraft lasted 2 hr. 9 min., six times longer than White's. Much to his surprise, however, he found his work programme too demanding. After manœuvring on the end of the tether, and performing other exercises, he had some difficulty in maintaining orientation; and when at length he went to the back of the spacecraft to put on the AMU back pack the exertion caused the suit temperature to rise and his helmet visor to fog. At this stage Cernan was having difficulty opening one of the arm rests of the AMU, and the astronaut was promptly instructed to return to the cabin.

Cernan, in company with other astronauts, had previously made a number of successful flights with the AMU in the cabin of a specially padded transport aircraft flown to induce weightlessness. The 166-lb unit, fitted into the back of Gemini's adapter module, required Cernan, wearing a 42-lb chest pack with a 25-ft umbilical tether, to ease his back into its form-fitting seat with the help of special hand rails on the spacecraft, and open the arm rests. He was then to detach the pack and launch himself into space, having first fixed a 100-ft extension to the tether.

The AMU, developed by Ling-Temco-Vought, was 32 in. high, 22 in. wide and 19 in. deep. Attached to the astronaut it had self-contained systems providing for life-support, communications, telemetry, propulsion, and manual and automatic stabilization. Its hydrogen peroxide propulsion system included twelve small nozzles mounted in the corners of the back pack which the astronaut worked by manipulating controls on the arm rests. The jets were designed to give him complete freedom of movement in space, forwards, backwards, and sideways.

The flight of Gemini 9 continued for the full 3 days, and on 6 June the craft made a perfectly controlled semi-ballistic re-entry, to land only 0·4 mile from the planned landing point some 345 miles east of Bermuda. In completing 46 revolutions of the Earth, the craft had travelled 1,255,630 miles.

The next Gemini mission was a complete proof of astronauts' ability to handle their craft in routine space missions, for it involved nothing less than a rendezvous with two orbiting vehicles. This included the Agena spacecraft which had been left in a parking orbit some 300 miles above the Earth since the curtailed Gemini 8 mission of 16 March.

A new Agena climbed successfully into orbit on the afternoon of 18 July, to be followed some 100 min. later by Gemini 10 with astronauts John W. Young and Michael Collins. The manned craft lifted off at 1720.26 hr. EST, its Titan booster placing it into an initial orbit of 100–167 miles. Rendezvous with the Agena at 185 miles altitude proceeded according to programme and after the two vehicles had successfully docked preparations were made to start the second manœuvre which would place the combined spacecraft in a transfer orbit reaching that of the Agena which had been in space four months. Collins's task was to climb out and retrieve the meteoroid collector pack. Use was made of the attached Agena's propulsion system to modify the orbit, and after disengaging their craft the astronauts manœuvred alongside the target vehicle.

After the cabin had been depressurized Collins emerged on the end of a 25-ft tether while Young held the spacecraft only a few feet away from the Agena. Collins floated across to the vehicle, removed the meteoroid collector, and after 30 min. returned to the Gemini cabin. It was the first time that any man had made direct personal contact with another orbiting object.

Once again 3 days were spent in space. The splash-down came at 1606.11 hr. EST on 21 July at a point 530 miles due east of Cape Canaveral, just 3 miles from the planned position. The craft completed 44 revolutions, covering a distance of 1,223,370 miles.

The next objective in the Gemini programme was to tie two spacecraft together in orbit so that experiments could be made in passive attitude stabilization. This was the task of Gemini 11 astronauts Charles Conrad, Jr and Richard F. Gordon, Jr whose vehicle left the launch pad at 0942.26 hr. EST on 12 September 1966. The spacecraft's initial orbit ranged between 100 and 177 miles, and as on previous occasions, an Agena target vehicle had gone up some 100 min. before. The object was a rendezvous during Gemini's first revolution of the Earth; also known as direct-ascent rendezvous, this was successfully achieved some 94 min. after lift-off. The astronauts used their on-board computer and radar equipment with only minimal assistance from ground tracking stations.

After the vehicles had docked, Gemini 11 used Agena's restartable rocket engine to extend the orbit in a wide ellipse ranging between 179 and 850 miles. Maximum speed in this orbit was calculated to be 17,967 m.p.h. Some of the most striking colour photographs of the Earth were taken on this mission, particularly views of India and Ceylon from 529 miles altitude and the north-west of Australia from near the orbital zenith.

These and other Gemini pictures placed geography in a new perspective and were closely studied by scientists concerned with cartography, geology, and oceanography for the wealth of detail they contained. So clear was one photograph taken over North Africa, on the second day of the flight, that a small black 'fuzz' on the Saudi Arabian desert was later identified as smoke from a fire which occurred at a crude-oil pipeline.

The astronauts docked and undocked four times with their Agena target and still had sufficient fuel left to make another unplanned rendezvous with the Agena.

After the cabin had been depressurized Gordon stood up for 2 hr. 8 min. with his head and shoulders projecting through the open hatch to photograph stars, clouds, and the Earth under different conditions of lighting.

The other period of EVA, planned to last 2 hr., had to be cut short. The craft were now back in a near-circular orbit at 184 miles and the astronaut's main task – during the thirty-first revolution – was to hook a 100-ft Dacron tether between the docked Agena and the docking bar on his own craft. The tether was stored in the Agena near the docking collar. Although the connection was achieved, Gordon was perspiring so heavily that moisture clouded his vision and overloaded the environmental control system of his pressure suit. He returned to the cabin after 44 min.

This was the second time that an EVA experiment had had to be cut short because of excessive astronaut exertions. To fix the tether Gordon had been forced to clamp his legs round the small end of Gemini in order to work in the weightless environment with two hands.

After the cabin had been resealed and the oxygen atmosphere restored to 5 p.s.i. pressure, Gemini 11 was separated from the Agena and the tether pulled taut, thus demonstrating how clusters of spacecraft might be fastened together in space to prevent them drifting apart. It also helped to assess a method of stabilizing vehicles towards the Earth by purely passive means, using the Earth's gravitational gradient.

The astronauts also showed how two spacecraft joined by a tether could be made to rotate round their common centre of mass to simulate gravity by centrifugal force. The crew fired Gemini's thrusters so that the craft began a cartwheel motion; and for the first time in space, a trace of artificial gravity was generated – about 0·00015 of normal Earth gravity – inside the Gemini cabin. This was a demonstration in miniature of the kind of technique that may have to be adopted in large space-stations of the future.

Despite the shortened 'space-walk', the mission was regarded as highly successful, and in returning to Earth the craft made the first automatic controlled computer-steered re-entry of a manned spacecraft. This was one of the most accurate Gemini landings. The craft splashed down 701 miles east of Miami, Florida, at 0859.34 hr. EST on 15 September only 1·5 miles from the planned position. It had travelled a total distance of 1,232,530 miles.

The last of the Gemini series was intended to clear up outstanding problems, particularly those affecting the ability of an astronaut to work outside his orbiting spacecraft. NASA had, in fact, entirely re-written the EVA programme in the light of prevailing difficulties which meant abandoning a second planned experiment with an AMU.

Gemini 12 was launched at 1546.33 hr. EST on 11 November 1966, 98 min. after its Agena target vehicle. The initial orbit ranged between 100 and 175 miles. Astronauts James Lovell and Edwin Aldrin made contact with Agena at 185 miles as planned and docked with it on their third orbit. A defect in Agena's primary propulsion system prevented a planned manœuvre into a 185–460-mile orbit. However, a retrograde burn of 43 ft/sec., using Agena's secondary propulsion system, changed the orbit of the combined vehicles to 160–177 miles, setting the orbit to phase with the 12 November total eclipse of the Sun over South America. Following a sleep period, there was another eclipse-phasing manœuvre which enabled the crew, through the Gemini windows, to obtain the first solar eclipse photographs from space.

The next task on that day was for Aldrin, the EVA astronaut, to take photographs of the Earth and stars with the cabin depressurized and the hatch open – as Cernan had done before his 'space-walk'. He had also to fix a portable hand rail on the outside of the craft. The hatch was open 2 hr. 29 min.

The big test was reserved for the third day. Aldrin climbed out of the cabin trailing a 30-ft umbilical tether. Using the hand rail previously mounted, he worked his way to the docked Agena. A pair of nylon body tethers, adjustable from 1·5 to 3 ft, were then fastened by means of pip pins inserted into holes at various points on the small end of Gemini and into rings on each side of the astronaut's parachute harness. After resting Aldrin was then able to perform his appointed tasks with little effort, pulling out the end of the 100-ft Dacron tether from the stowage near Agena's docking collar and looping the end over Gemini's docking bar. He also exposed a micrometeoroid collector plate on the Agena.

Later, the astronaut was to transfer to the spacecraft's adapter module, locating his feet in 'overshoe' restraints. After another rest period he began a 'work task evaluation' at a 30 in. by 30 in.

panel on which were mounted electrical and fluid connections, combinations of hooks and rings, strips of Velcro, and fixed and removable bolts. First he worked – loosening and tightening bolts, plugging and unplugging sockets – using only the foot restraints and then with only the body tethers.

After this exercise Aldrin made his way back to the spacecraft's nose to perform another sequence of work-tasks on a smaller panel, using the same body tethers as before to help control his body position. Before working at each site he fixed a camera so that his subsequent actions could be fully recorded.

Aldrin took about a dozen 2-min. rest periods and making full use of restraint straps and hand-holds, he was far more successful than previous astronauts in resisting fatigue. His pressure suit did not overheat, and at no time did his heart-rate exceed 130 per min. He remained outside the cabin for a record time of 2 hr. 29 min. 25 sec.

After Aldrin had returned the spacecraft were undocked and allowed to drift at random. The craft stabilized in vertical formation after 1 orbit, the tether exercise lasting a total of 4 hr. 17 min. Then, on the last day of the flight, Aldrin performed another 52 min. of open-hatch photography. This brought his total exposure to space during the three EVA periods to over 5 hr. 30 min.

The mission was rounded off by an accurate return on 15 November ending with a splash-down at 1421.04 hr. EST, 720 miles south-east of Cape Canaveral. The craft had covered 1,628,510 miles, making over 59 revolutions of the Earth. This landing only 2·7 miles from the aiming point brought an historic series of spaceflights to a worthy close. The entire spectacle, including landing, helicopter pick-up of the astronauts, and their enthusiastic welcome aboard the aircraft carrier *Wasp*, was televised live via the Early Bird satellite and relayed throughout the United States.

At a ceremony at his Texas ranch on 23 November, President Johnson presented the Gemini 12 astronauts with Exceptional Service Awards of the National Aeronautics and Space Administration. These acknowledged not only Lovell and Aldrin. They honoured all the astronauts (fifty were then in training), the check-out and launch crews; the flight controllers; the tracking station personnel; recovery forces, and the men and women who designed and built the spacecraft, the Titan, Atlas, and Agena rockets, and the vast array of supporting equipment.

The record was impressive. America was now far ahead of the Soviet Union in almost every aspect of manned spaceflight. In terms of distance travelled, Mercury and Gemini had logged 17,616,000 miles; Vostok and Voskhod 7,471,000 miles. But still more impressive was the fact that the two Powers combined had performed 25,000,000 miles of pioneering space travel apparently without loss or injury to any of the participants.

The bridge to the Moon was fast being erected, and the Apollo programme upon which all effort now concentrated seemed capable of achieving John Kennedy's goal of placing American astronauts on the lunar surface by 1970.

The stage had been reached when three Apollo spacecraft had been launched unmanned by the Uprated Saturn 1 substituting for the larger Saturn 5 of the actual Moon programme. The first unmanned Apollo sub-orbital launching, on 26 February 1966, was mainly to prove re-entry characteristics of the command module and by the end of the year NASA was approaching the stage where the first Apollo command and service modules would be tested with a three-man crew in Earth-orbit. The flight, scheduled to begin on 21 February 1967, was expected to last a maximum of 14 days.

The first spacecraft disaster, when it struck, therefore was all the more stunning for happening not in space but on the ground. It removed from the family of space pioneers three brave men, two of whom had already made space history. They were Virgil Grissom, the Apollo command pilot, who had previously made spaceflights in 1961 and 1965, and Edward White, the first American to leave a spacecraft in orbit; the third was Roger Chaffee who had yet to make his mark in space.

The disaster occurred on 27 January 1967 at Launch Complex 34 at Cape Canaveral as the astronauts were participating in a practice countdown in readiness for the first manned orbital flight aboard the Apollo 204 spacecraft. The command and service modules were mounted in the nose of the Uprated Saturn 1 launch vehicle, which was unfuelled, and the astronauts in their cabin had already been there over 5 hr. The service tower was in place round the vehicle and at the time of the tragedy the spacecraft was operating on external power.

The first warning of fire came at 1831.04 hr. EST while the countdown was holding at T—10 min. (i.e. 10 min. from the time of simulated lift-off). Up to this point NASA reported only minor difficulties with the equipment, the reason for the 'hold' being to provide an opportunity to improve communications between the spacecraft and the ground crew. Cabin pressure, cabin temperature, and oxygen suit supply temperature were normal. Cabin pressure in the pure oxygen environment was about 16 p.s.i. before the fire – a value necessary at ground-level to achieve the desired pressure balance, but which in an actual launching would be reduced to about 5 p.s.i. as the craft climbed into space.

The fire appeared to start unseen at the lower left of the capsule below the feet of astronaut Grissom. The investigating Board thought it possible that a 'hot wire' or short-circuit initiated a fire that spread to nylon netting (used to prevent loose objects from floating into equipment crevices while in zero-g), Velcro fastening material (used to attach equipment to the cabin interior), and the environmental control unit insulation.

It took some 15 sec. from the first report of fire (by Chaffee) until the cabin shell burnt through between the inner and outer hulls. Pressure inside the capsule immediately prior to this was 36 p.s.i. All three suits burned through, Grissom's (on the extreme left of the cabin) receiving the greatest exposure to flame and Chaffee's (on the opposite side) the least. Yet the fire had surprising variability in intensity and timing. Dr Robert C. Seamans, Jr, Deputy Administrator of NASA, cited the example of an aluminium tubing handle which had a hole burned through it, indicating a temperature at that point of at least 760° C, yet its nylon hinge within 2 in. of the melted spot was relatively undamaged indicating a temperature there of less than 254° C.

Dr Seamans reported:

> One hypothesis, supported by the cabin pressure history, assumes a small, low-grade fire whose heat was at first largely absorbed by the spacecraft structure and that was burning at the time of the first crew report; that fire may have continued for as long as 10 sec. A more intense fire may have developed, causing the rapid increase in cabin pressure. This fire was probably then extinguished by the depletion of oxygen.

Death was attributed to asphyxiation caused by inhalation of smoke.

Grissom had recorded an epitaph for himself and his fellow-astronauts in an interview which followed his epic mission in Gemini 3: 'If we die,' he said, 'we want people to accept it. We are in a risky business and we hope that if anything happens to us it will not delay the programme. The conquest of space is worth the risk of life.'

Almost immediately NASA announced the names of the replacement crew: Captain Walter M. Schirra, Jr, command pilot; Major Donn F. Eisele, senior pilot; and Major Walter Cunningham, pilot. This was the back-up team whose training had paralleled that of the dead astronauts. Later, when the extent of the modifications required to ensure safety of the Apollo spacecraft was fully realized, active training of crews for Apollo missions was temporarily suspended and the astronauts previously chosen to make them (including the reserve crew of AS-204) were stood down. It was some time before American astronauts were ready to go into space again.

Thus began the era of human exploration of the Moon, on the wings of both triumph and tragedy.*

* Two trainee astronauts, the former test-pilot Elliot M. See, Jr., and Captain Charles Bassett of the US Air Force, had been killed on 28 February 1966 when their T-38 Talon jet-trainer crashed into the McDonnell factory in St Louis. They were preparing for the three-day Gemini 9 mission subsequently flown by Thomas Stafford and Eugene Cernan. Elliot See was to have been command pilot and Charles Bassett was being trained for a 90-minute 'space-walk'. Another trainee astronaut, Theodore Freeman, was lost in 1964 when a goose smashed through the windscreen of his T-38.

CHAPTER THREE

PROJECT APOLLO

The spaceport for America's Apollo Moon programme was built on a wasteland of sand and swamp adjacent to Cape Canaveral on Merritt Island. To achieve it NASA spent 1,000,000,000 dollars in what must be the biggest and most expensive construction project of all time. The major facility, Launch Complex 39, required 1,000,000 tons of steel and nearly 17,000,000 tons of concrete. The site occupies 80,000 acres and is criss-crossed with 100 miles of roads; and dominating the skyline is one particular structure known as the VAB or Vehicle Assembly Building.

The VAB was the assembly hangar for Saturn 5 and its Apollo spacecraft, and so gigantic are its proportions that no fewer than four of these massive vehicle combinations could have been vertically assembled at the same time. Each assembly bay has work platforms which can be extended vertically and horizontally to provide access to a launch vehicle at any level. Next door is the Launch Control Center and a full simulated countdown can be carried out while the vehicle is still inside the building. Nearly 600 ft long, 418 ft high, and 410 ft wide, the building is so large that, were it not for the special air-conditioning system, clouds could form inside.

A lunar mission actually began in the VAB for it was here that the complete vehicle was assembled on a mobile launch platform and fully checked-out ready for flight. After final inspection, a huge door in the VAB was opened and the unfuelled 364-ft vehicle was slowly moved out complete with launch platform and 380-ft umbilical tower on a gigantic diesel-powered crawler-transporter. This lethargic monster with its 12,000,000-lb load moved at only 1 m.p.h. along the crawlerway which led to the launch pads some 3 miles distant. At its destination were two launch complexes. Each had an elevated concrete and steel structure at its centre which received the platform-mounted vehicle and umbilical tower on the mobile transporter after it had negotiated a ramp.

After the rocket and its mobile launch platform were in position on the support pedestal, technicians gained access to the vehicle by means of hinged walkways which extended from the umbilical tower. A 400-ft mobile service tower was also moved into position close to the rocket. Basic systems were now re-checked and a number of vital interconnections made between launch pad and vehicle. At length the spacecraft was fuelled. Tanking the rocket with kerosene (RP-1), liquid oxygen, and liquid hydrogen occurred last. Finally, the flame deflector was moved into position on rails in the trench beneath the launch pedestal. Its mobility allowed use of a second flame deflector when the first was being refurbished after a firing.

The three astronauts, of course, participated in the last hours of the check procedure. Just before launching, the service tower was moved back while the umbilical tower was left in position. As the countdown reached its climax the five powerful F-1 engines in the first stage ignited, and ponderously the 6,000,000-lb space leviathan lifted from the pad. The engines were consuming kerosene and liquid oxygen at the prodigious rate of 5,000 gallons per sec.; thrust exceeded 7,500,000 lb. The noise was deafening.

In the Launch Control Center 3 miles away a vast array of electronic monitoring equipment registered every vital part of the launch procedure as technicians watched the lift-off on television screens.

In the first 2 min. the rocket burned 2,100 tons of propellant. After the vehicle had made its programmed turn over the Atlantic Ocean, and was climbing at a height of some 30 miles, the S-IC first stage jettisoned leaving the S-II stage to continue propulsion. At 2·54 min. from lift-off, the launch escape system was jettisoned. When the vehicle had reached 100 miles some 9·04 min. into the flight, the second stage fell away; and the S-IVB third stage began thrusting to place the spacecraft into a parking orbit close to the Earth. Speed was 17,500 m.p.h.; elapsed time 11 min. 52 sec.

This was Apollo's first 'breathing-space' in the flight programme, and the time when systems could be re-checked in conjunction with ground control. At NASA's Johnson Space Center at Houston, Texas, computers calculated Apollo's orbit and determined the time when the S-IVB's engines must be restarted to enter the spacecraft on the desired transfer orbit to the Moon.

If all was well, at the appointed time, the third-stage engine re-fired and the astronauts prepared for the next vital stage of their lunar journey. If translunar injection was made on the second orbit, the time into the mission was just over 3 hr. The speed at which engine cut-off occurred was critical. It had to be between 35,000 and 36,500 ft/sec.

Soon after the vehicle had entered the coast phase of the mission preparations were made to start the transposition and docking manœuvre which prepared the craft for the subsequent lunar landing procedure.

This entailed separating the command and service modules from the expended S-IVB rocket stage and blowing free the adapter panels which shrouded the lunar module carried in this rear compartment. The manœuvre was completed by flying the command and service modules out a short distance ahead, turning them through 180° by use of gas-jet controls, and gently bringing them back so that the docking mechanism on the nose of the command module entered the docking drogue of the lunar module attached to the S-IVB.

With this achieved the lunar module was pulled away from the rocket stage which was left behind.

Mid-course corrections were then carried out to align the trajectory accurately with the Moon which itself was moving relative to the Earth at about 2,300 m.p.h. The spacecraft, of course, did not maintain a uniform speed as it coasted towards the Moon; speed varied constantly under the influence of gravitation, averaging about 5,500 m.p.h.

To ensure accuracy and timing, the necessary velocity adjustments were computed on Earth and relayed to the spacecraft. Communications were maintained by a chain of stations round the world, including the NASA Deep Space Network Stations in the United States, Spain, South Africa, and Australia. As one station disappears below the horizon on the rotating Earth, another appeared to keep the link open. The DSN Stations have highly directional 85-ft dish antennae providing a range accuracy of 49 ft at the distance of the Moon; they have been used regularly to receive data from space probes including television pictures of the lunar surface. The largest directional station at Goldstone has a 210-ft dish with a range two and a half times that of the other antennae; it is capable of following a spacecraft to the edge of the Solar System.

Mid-course corrections were made after Apollo had been orientated at the prescribed angle by firing the service module's rocket engine in an accurately controlled burst. In a typical mission as many as three velocity adjustments might be necessary. A first 'coarse' correction could be applied as early as 5 hr. 6 min. into the flight; a second smaller adjustment might be made at 55 hr. 30 min. and a third at 63 hr. 15 min.

During the 70 hr. of flight the astronauts had many tasks to perform working regular shifts. They were in constant touch with Houston by radio and television.

Apollo carried fuel-cells for electrical power supply; liquid hydrogen and oxygen flowed from separate tanks into the fuel-cell battery, water being produced as a by-product. A similar system was used in long-duration Gemini missions. In Apollo the water was used for drinking and general cleansing purposes. Food included the hydrated kind tested during the Gemini programme. Rest and sleep were taken at regular intervals and spacesuits could be removed in the cabin's shirt-sleeve environment.

One of the principal tasks during the translunar journey was that of checking equipment in the lunar module. The action of docking had previously made a pressure-tight seal between the two vehicles. In order to pass freely between them, however, the astronauts had first to open the hatch covers and remove the docking fixture. The interconnecting tunnel was then free of obstruction and transfer between them simply a matter of crawling through. The lunar module was, of course, pressurized at the same value as the command module. The astronauts emerged into the lunar module through a hatch in the roof of the compartment aft of the control cabin. After ensuring that everything was in order they rejoined their companion in the command module.

After the final mid-course correction the Moon would be

looming large in the sky exerting increasing gravitational pull and tending to speed up the vehicles. Again computers on Earth would monitor equipment on board the spacecraft to ensure that the trajectory was exactly right to achieve a 'miss' of 90 to 100 miles. Normally insertion involved a two-stage firing of the service propulsion engine, the first to put the mooncraft into an elliptical orbit of about 113 by 314 km. (70 by 195 miles), and the second to circularize the orbit at about 111 km. (69 miles). The spacecraft would be approaching the Moon in a backward attitude, the large nozzle of its service module facing forward in the path of flight. To place the craft into orbit round the Moon, this engine had to be used as a retro-rocket to slow down the vehicle near the point of closest approach. Elapsed time at the point of retro-fire could be approximately 64 hr. 15 min.

When retro-thrust had been applied – and following velocity adjustments – the combined craft began to swing into orbit round the Moon some 69 miles above its surface. The astronauts now looked down upon the vast airless panorama of dry lava 'seas', craters, and mountains, already familiar from the reconnaissance pictures taken by unmanned Ranger, Orbiter, and Luna space probes. Sunlight flooded the surface casting intense black shadows. In the absence of a protective blanket of air, meteoroids and radiation at all frequencies bombarded the surface.

When the Sun was directly overhead temperatures could exceed 100°C. A lunar day and a lunar night are approximately 14 Earth days long, and as the night shadow slowly crept across the Moon's face, in the absence of an absorbing atmosphere, there was a sudden fall to around −160° C.

The spacecraft, orbiting at about 3,500 m.p.h., completed 1 revolution of the Moon every 2 hr. During the first part of the orbit the astronauts made equipment checks, sighted landmarks for orbit determination, and updated the guidance navigation and control sub-systems. Now final preparations could be made for the task of landing the lunar module at the appointed site near the lunar equator.

After transferring to the lunar module, the commander and LM pilot checked all vital equipment including propulsion, guidance control, and 'abort' systems. Communication with Earth was possible when the vehicle had emerged from behind the Moon which would otherwise block the reception of radio signals. If everything was in order the 'go' signal was received from the Johnson Space Center in Houston, and the landing procedure began.

The commander of the lunar module and the systems engineer took up their positions in the lunar module. They did not have seats or crew couches, but stood at the control consoles secured by restraint harnesses.

Two actions remained to be performed before casting off. One was to extend the landing legs of the lunar module. Another was to replace the docking mechanism in the nose of the command module ready for the return lunar-orbit rendezvous and docking. The hatches to the interconnecting tunnel in both vehicles had also to be closed to form gas-tight seals.

After the astronauts had brought all systems to operational readiness and assuming there were no snags, they undocked the landing craft from the Apollo parent. The two vehicles moving in close formation above the Moon were in touch by radio, and before

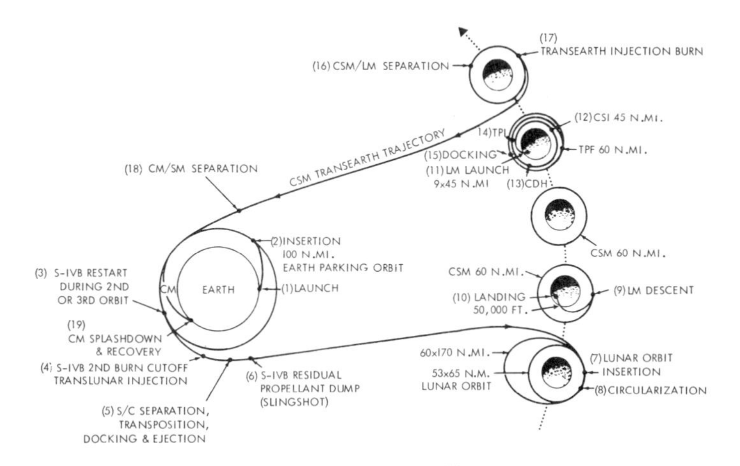

Apollo 11 flight profile

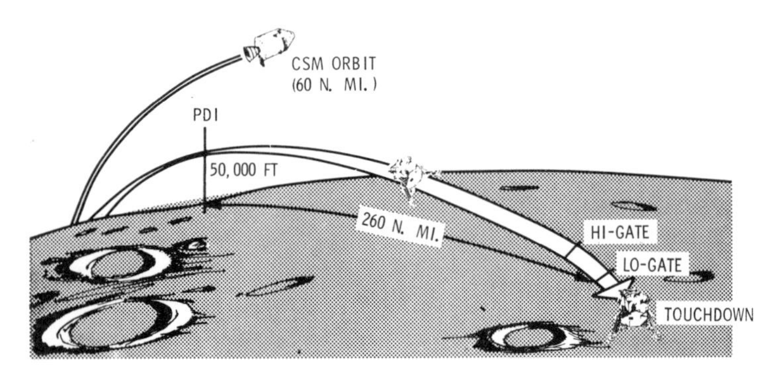

Phases of powered descent

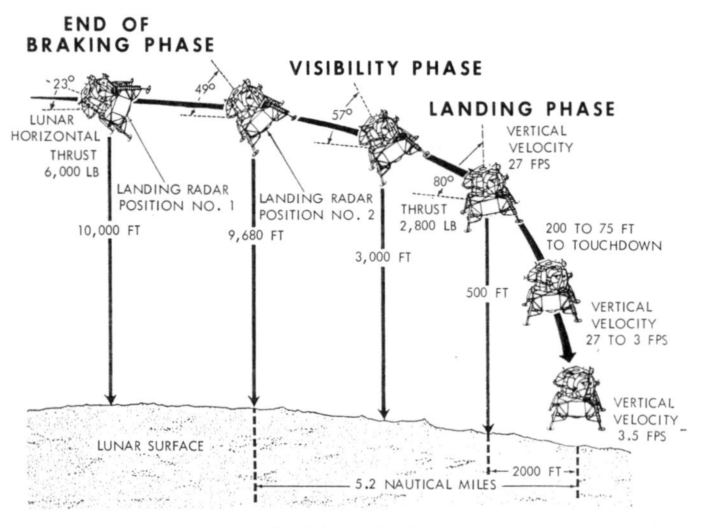

Final descent path

starting the descent the lunar module astronauts conducted test-manœuvres to ensure that the thrusters of the Reaction Control Sub-system (RCS) were in working order. Each phase of the test procedure was monitored by the third crew-member (command module pilot), who remained behind in the parent vehicle.

Landings were made ideally within a few hundred yards of the chosen site, based on accurate maps prepared from the thousands of close-up photographs obtained principally by Lunar Orbiter probes. Once again orbital parameters necessary to ensure an accurate descent trajectory were computed on Earth and stored in the spacecraft.

With the lunar module correctly orientated for retro-fire, its rocket engine reduced speed sufficiently for the vehicle to enter an elliptical orbit. If no further action had been taken this would have caused it to swing within 50,000 ft of the Moon's surface before returning to the original apogee on the opposite side of the Moon.*

However, if all proceeded normally, the point of closest approach was approximately 225 miles uprange of the proposed landing site.

As the lunar module's elliptical orbit had a period similar to that of the parent vehicle, the two craft remained in line-of-sight communication and the approach path of the landing craft could be verified by mutual radar tracking from both vehicles.

Retro-fire for landing normally began at the time of closest approach, using near-maximum thrust along a minimum-energy trajectory. Engine thrust was variable between 8,200 and 1,100 lb. Owing to the high pitch angles required for the braking manœuvre, however, the landing site was not immediately visible to the astronauts and the descent was performed automatically to within 700 ft of the surface. The manœuvre proceeded in three stages. First the *braking phase* from approximately 50,000 to 10,000 ft; second a *final approach phase* from approximately 10,000 to 700 ft; and third the *landing phase* which terminated at touchdown.

The lunar module was orientated ready for the final approach phase some 2 min. before reaching the 10,000-ft mark and then the astronauts had an unobstructed forward view. At that time

* A modified procedure in which the LM and parent remained together.

new data could be fed into the guidance system to select an alternative landing site if the approach path to the primary area was not exactly right.

Once rocket thrust counterbalanced the spacecraft's weight (now approximately $\frac{1}{6}$th of the Earth value), the vehicle was able to hover, ascend, or move sideways in the low-gravity environment. Even at this stage, if anything was wrong, the mission could be 'aborted' by firing off the ascent stage of the spacecraft which formed an escape capsule capable of returning to lunar orbit. However, if all was well, the craft could continue in the hovering-drift mode for up to 2 min. while the astronauts, peering down at the lunar surface through their windows, chose a suitably smooth landing place. Although the craft could land entirely automatically using a radio-altimeter to gauge height, small rocks or other obstructions were best judged by the astronauts themselves. The standard procedure, therefore, was for the landing to be performed by the command pilot. The method of manual control was similar to that of a helicopter. The astronaut worked a translation controller with the left hand and an attitude controller with the right. Control actions, transmitted electrically, worked valves in the Reaction Control Sub-system which released propellant to the thrusters in appropriate measure; for example, tilting the vehicle in the hover mode produced forward or backward translation.

The procedure was well rehearsed on Earth where the astronauts learnt to control jet-powered simulators. The shadow cast by the spacecraft itself was a useful gauge of height over the surface and when this was down to 3 ft, engine thrust was cut and the vehicle dropped to rest on the landing legs.

With landing safely accomplished, the first task was to check all systems in readiness for the return take-off, for if anything was wrong the maximum time had to be allowed for rectification. A major engine fault could leave the astronauts stranded on the Moon. A full report was made both to the third astronaut in lunar orbit on VHF and directly to Earth using S-band frequencies (employed in Apollo for purposes of both tracking and communications). This decided if the planned stay-time could be accomplished.

With check-out completed the exploration phase began. It was no longer a rule that only one astronaut at a time emerged from the ship, while the other remained inside to monitor operations and maintain communications. From Apollo 11 it had been possible for both crew-members to make joint excursions on the lunar surface. Beneath his EVA pressure suit the lunar explorer wore next to the skin a water-cooled undergarment which served to control internal temperature. On his back was fitted a portable life-support system, which supplied cool purified oxygen to the suit at 4·8 to 5·2 p.s.i. There were also special gloves and thick-soled insulated boots.

To leave the craft the lunar module's cabin was depressurized while both astronauts retained suit pressure. The forward hatch beneath the centre instrument consoles was opened and the explorer-astronauts climbed out using the platform and stairway that extended from the front.

First task for the men outside was to inspect the craft for damage such as might be caused by small rocks thrown up by the engine exhaust. Next they extended the S-band antenna so that photographs from a small hand-held television camera could be sent back to Earth via the S-band link.

The portable life-support system allowed the astronauts to work for periods of up to 3 hr.* Apart from taking photographs, they collected samples of lunar rock and made observations and measurements of the lunar topography. All the time they were in direct voice contact, over VHF, with Houston Mission Control.

After their time of exploration was over, the astronauts returned to the lunar module. The hatch was closed and pressure restored. Cabin repressurization took about 1 min.

The portable life-support system was now replenished while voice reports were transmitted to Earth together with scientific data. Further excursions outside the craft could be made using the same procedure.

Scientific experiments, of course, were an important feature of the exploration programme. To facilitate this effort NASA provided Apollo Lunar Surface Experiments Package (ALSEP)

* Later, extended.

payloads, which were entirely self-contained ready for the astronaut to remove from a compartment in the descent stage between two of the landing legs.

As sunlight to power solar cells was not available during the lunar night, and large quantities of chemical energy would be required for long-term fuel-cell operation, ALSEP packages were powered by a SNAP-27* thermoelectric nuclear generator. Using plutonium-238 as the heat source the unit supplied 56 W of electricity for 1 year but was in fact designed to last much longer.

Fuelling the generator was a task in itself. The plutonium fuel was carried in a storage cask separately attached to the side of the descent module.† First the astronaut removed the two ALSEP packages and the inert SNAP-27 generator. To fuel the generator the astronaut first hinged the storage cask into a horizontal position and extracted the fuel capsule with a special handling tool. With the same tool he inserted the capsule vertically into the generator housing and locked it in position.

The two ALSEP packages were then assembled on a bar and carried by the astronaut to the desired location. There was no difficulty in transporting the system over several hundred feet since the 180-lb load under lunar gravity conditions was only about 30 lb. It was important to set the equipment some distance from the lunar module to avoid possible damage from flying debris as the vehicle took off.

At the experiment site the generator and scientific equipment, linked by cables to a telemetry transmitter, were laid out around a central platform. Apart from serving as a junction-box for the wiring, the platform carried the helical telemetry antenna which was gimbal-mounted at the top of a short mast. To achieve the desired signal strength at the receiving station the antenna had to be carefully aimed at the Earth.

After the astronauts had activated the instruments they con-

* SNAP is the contraction of Systems of Nuclear Auxiliary Power.

† This mounting arrangement was necessary in order to prevent the radioisotope fuel from dispersing on the Earth in the event of a launch mishap. The storage cask, held by metal bands, had a heat shield and was designed to re-enter as a stable atmospheric vehicle.

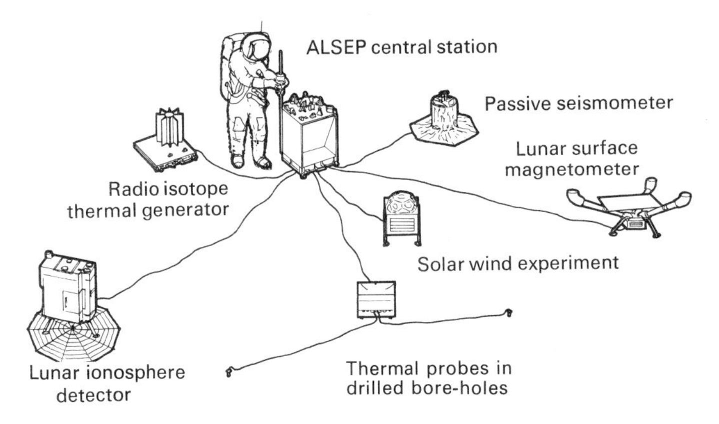

Apollo Lunar Surface Experiments Package (ALSEP).

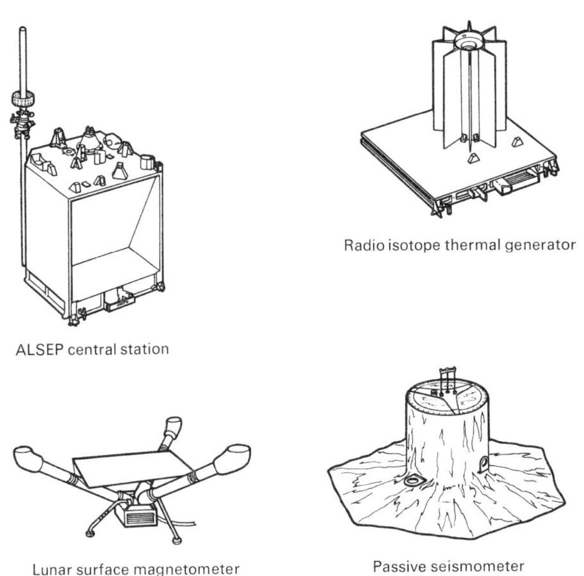

Some ALSEP instruments.

tinued to supply data from the lunar surface long after the men had returned to Earth.

The first seven ALSEP experiments planned by NASA were as follows:

1. *Passive Lunar Seismic Experiment.* A 3-axis seismometer to measure lunar tremors or 'moonquakes' allowing study of the Moon's interior to its centre, i.e. whether it has a crust and core and whether it is layered in structure.
2. *Lunar Tri-Axis Magnetometer.* This instrument, similar to ones flown in unmanned spacecraft, to measure the Moon's internal magnetic field as well as the interaction of the solar wind with the magnetic field round the Moon.
3. *Medium Energy Solar Wind Experiment.* A plasma spectrometer to measure the velocity and direction of protons, electrons, and alpha-particles in the solar wind as they arrived at the Moon and the interaction of these particles with the lunar surface.
4. *Suprathermal Ion Detector.* This experiment measured the Moon's ionosphere by sampling ions in a wide range of energies to determine how strongly it is affected by the solar wind.
5. *Lunar Heat Flow Measurements.* This instrument measured the outflow of heat from the Moon's interior through the surface to provide information on the distribution of radioactive elements and the thermal history of the Moon, including volcanism.
6. *Low Energy Solar Wind.* As in the case of experiment 3, this instrument studied solar wind particles, but in the lower energy ranges.
7. *Active Lunar Seismic Experiment.* After the instrument was activated, an astronaut hit the lunar surface with a thumping device as he walked out to 1,000 ft from the lunar module. Beyond that distance, a small mortar device was used to fire small projectiles to land on the surface; the instrument measured the local tremors to obtain information on physical properties of the lunar crust to a depth of about 5,000 ft.

Following the period of lunar exploration, the astronauts sealed the hatch of the lunar module for the last time and repressurized

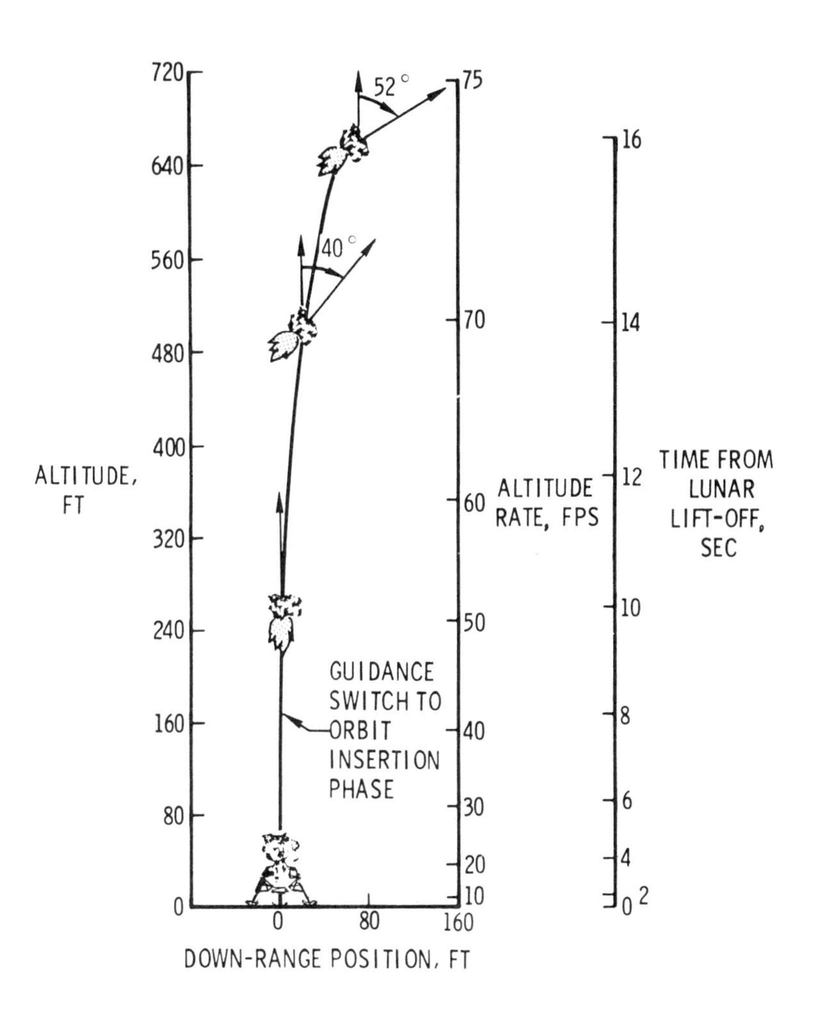

Lift-off from Moon

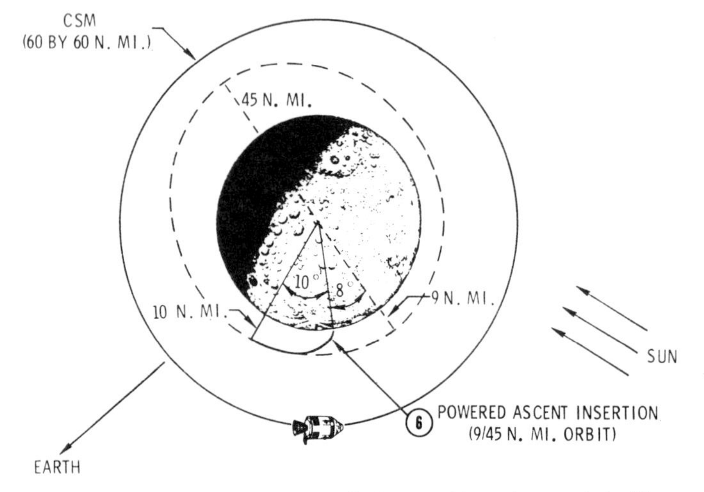

Insertion of LM ascent stage into preliminary parking orbit round the Moon before docking with CSM.

the cabin. All systems were brought to operational status and the vehicle's launch readiness confirmed to Apollo Control on Earth and also to the astronaut orbiting in the parent spacecraft.

The launch method, originally developed in feasibility studies by the British Interplanetary Society between 1938 and 1947, involved using the spacecraft's leg-supported descent stage as the launch platform. Not only did this provide a stable platform for the lift-off but leaving the base section behind considerably reduced the deadweight to be lifted back into lunar orbit.

The ascent path had to be precisely matched with the orbit azimuth and approach time of the parent spacecraft. The situation recalled the prior work with Gemini and Agena in which the aim was to rendezvous during the first orbit. Acting on computed data stored in the lunar module, the astronauts could have manual control over engine ignition or programmed command.

Lift-off followed a countdown when the CSM was approaching in line-of-sight over the horizon. The ascent stage ascended on a thrust of 3,500 lb, its engine burning continuously up to the point of insertion into the transfer orbit which brought it to the rendezvous position. During the first 12 sec. the craft rose vertically; then followed two 'pitchover' phases, the first steeper than the second. The engine was cut off at a height of some 50,000 ft approximately 7 min. after lift-off. At this point the vehicle was in an ascent transfer orbit, which nominally intercepted the Apollo parent at the first intersection of the two vehicles' orbits.

If, for any reason, lift-off of the lunar module had to be delayed for more than 1·5 min., the craft could 'loiter' in the 50,000-ft altitude parking orbit before a second engine firing was made to insert the craft into a revised orbit for rendezvous. Lift-off from the Moon could not be left longer than 8 min. as in this time the parent vehicle would have disappeared below the Moon's horizon. In this event, the astronauts would have to wait some 2 hr. for the next launch opportunity on the succeeding orbit.

However, if the craft had been successfully injected into the transfer orbit for rendezvous, it should approach the Apollo parent within 500 ft. The commander manually manœuvred the lunar module to a docking attitude and increased or decreased the rate of closure until docking was accomplished.

Although the parent vehicle normally remained passive during this operation, if necessary it could perform rendezvous and docking. In fact, should the lunar module fail to achieve rendezvous there were contingency plans whereby the Apollo parent would modify its orbit to 'chase' the smaller craft and dock with it. Reserve propellant was carried for this purpose.

With docking completed the two astronauts prepared to rejoin their companion in the command module. Pressures between the two craft were equalized, systems of the ascent stage turned off, and after removal of the docking mechanism in the interconnecting tunnel, scientific equipment and lunar samples were transferred to the parent vehicle.

With the three astronauts reunited, the docking fixture was left in the lunar module, and the cabin hatch resealed. The lunar module was then jettisoned.

After more equipment checks, preparations were made to restart the engine of the Apollo service module for the return flight. Once again the transfer orbit and launch timing were computed on Earth and stored in the spacecraft. At the appropriate moment the 21,500-lb thrust engine fired propelling the craft to a speed of about 5,600 m.p.h. to achieve escape from the Moon. The return journey lasted approximately 60 hr. As Apollo approached the Earth, speed increased to nearly 25,000 m.p.h. because of the Earth's gravitational pull.

Ground stations again supplied data for mid-course manœuvres, for accuracy of the trans-earth trajectory was vital to the astronauts' safe return. Communications were assisted by the Atlantic satellite in geo-stationary orbit, which passed signals received from the spacecraft between the NASA ground station on Ascension Island and the United States as the vehicle headed towards its destination.

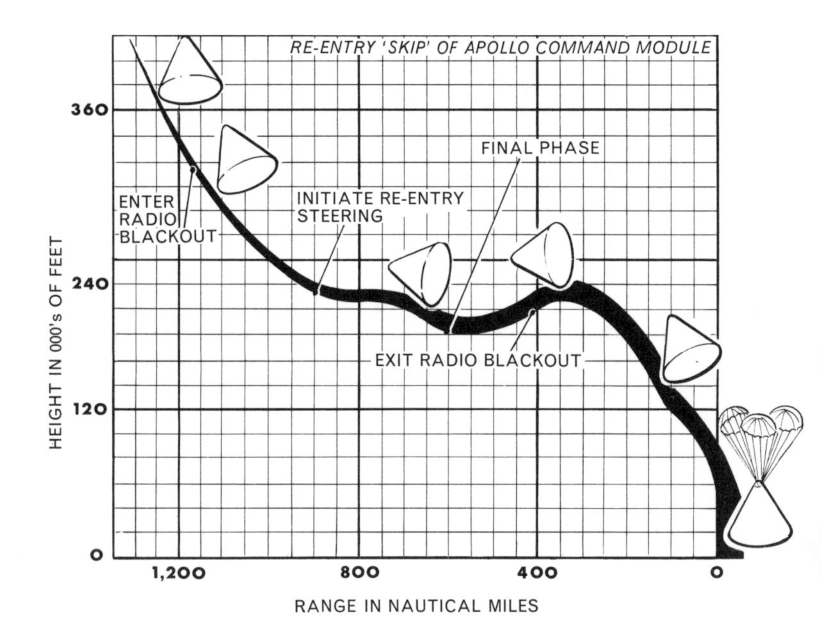

Apollo CM re-entry sequence: altitude versus distance to splashdown.

The approach trajectory had to hit a re-entry corridor only 26 miles wide. Otherwise the vehicle would be in danger of missing the Earth and being deflected into a remote part of space. On the other hand, dipping too steeply into the Earth's atmosphere could mean destruction of the spacecraft by excessive frictional heating.

Thus, the mid-course manœuvres had to be carried out with exceptional accuracy. Then, with the vehicle on target, and about 15 min. before entry into the Earth's atmosphere, the service module was jettisoned. This left the conical command module to make the final descent.

Using pitch control jets the spacecraft commander turned the vehicle over so that the blunt heat shield faced forward to absorb re-entry heating. There were no retro-rockets to brake the speed. As the craft cut through the atmosphere temperatures on the shield's surface could reach 2742° C.

It was not entirely a ballistic return. The command module, with its offset centre of gravity, could produce a certain amount of aerodynamic lift. The ablative material protecting the craft was thickened on one side to provide for this. In this way the commander could manœuvre the capsule to stay in the narrow re-entry corridor and select his landing point. He could, in fact, guide the craft to a selected landing area at sea as far as 5,000 miles away.

By the time the craft had dropped to within 30,000 ft, drag-braking had reduced the speed to around sonic velocity. At 25,000 ft, drogue parachutes were deployed, followed at 15,000 ft by three huge ringsail landing parachutes which lowered the capsule gently into the sea. As the craft splashed down the parachutes were severed and the astronauts awaited helicopter rescue.

CHAPTER FOUR

APOLLO DIARY

A large part of the civilized world became one family in July 1969 when Man first landed on the Moon. A million people watched the blast-off of the Apollo 11 moonship at Cape Canaveral and television networks beamed abroad, via satellite, telecasts totalling 230 hr. to reach an additional audience of some 500 million.

No one could know what the outcome would be, or indeed if the explorers would ever return. Who indeed could forget those anxious moments of the first lunar touchdown, or the misty image of Neil Armstrong slowly descending the ladder of the 'Eagle' lunar module before vaulting lightly on to the virgin moondust?

However, to spotlight the triumph of Apollo 11 is to ignore essential preliminaries. The Apollo 204 fire tragedy of 1967 had cast gloom over the entire space community and men and women of NASA and the associated industries worked night and day to implement the recommendations of the accident inquiry (pages 279–80).

First of all flights were made unmanned to test the Saturn 5 moon rocket and the lunar module (see table page 288).

Manned flights of the modified CSM began with Apollo 7 when astronauts Walter Schirra, Donn Eisele, and Walter Cunningham orbited the Earth in an Apollo command module for 10 days beginning on 11 October 1968.

With the Russians making final unmanned tests of their Zond circumlunar spacecraft, NASA was under pressure to achieve the next major step of a flight around the Moon before the Russians could fly a man out and back on a free-return trajectory. They were still, however, far from the objective clearly stated by Konstantin Feoktistov in a 1965 *Tass* interview – the 'assembly of equipment in space' and 'the development of facilities for landing an expedition on the Moon and ensuring its return to Earth'.

The decision was taken to bring forward the next Apollo

mission – originally meant to test the lunar module in Earth-orbit – and head American astronauts, for the first time, in the direction of the Moon. In many eyes this involved unacceptable risks. Not only would the mission follow just two unmanned test flights of the mighty Saturn 5 rocket but NASA insisted that the lunar goal of Apollo 8 should be orbital. If Apollo's engine failed to re-ignite at the Moon end of the journey the astronauts would be marooned beyond recall. There was no way in which they could be rescued.

As it turned out Apollo 8 was a textbook mission writing large the names of Frank Borman, James Lovell, and William Anders in the history of lunar exploration.

Flown without a lunar module their mission completed 10 orbits of the Moon on Christmas Eve and Christmas Day 1968 giving television viewers back home a breathtaking panorama of the Moon's desolate surface from a distance of just 111 km. (69 miles). That men were behind the camera nearly a quarter of a million miles from Earth added an extra dimension of excitement.

For the Soviet Union it was the breaking point. Struggling to achieve the one major 'first' of sending a man *around* the Moon on a free-return trajectory (but capable of achieving little more), the Russians were eclipsed by the sheer audacity of the Apollo 8 venture.

There followed two more manned Apollo flights in anticipation of the first lunar landing. In Apollo 9 James McDivitt, David Scott, and Russell Schweickart tested in Earth-orbit the full Apollo spacecraft and checked rendezvous and docking procedures. In a final dress rehearsal Apollo 10 with astronauts Thomas Stafford, John Young, and Eugene Cernan tested the complete hardware in orbit around the Moon. During this latter mission a manned lunar module was separated from the Apollo CSM in full simulation of a landing mission but without insertion of the LM into the final descent trajectory. The closest the astronauts got to the surface was about 14,325 m. (47,000 ft).

On 16 July 1969 the great day of the moonshot itself finally dawned clear and bright over the Kennedy Space Center.

It was a day that left the world breathless, anxious and united in a common prayer for the safety of three brave men in a frail machine who could so easily be lost.

At 0932 hr. local time watchers saw the bright orange flame burst from the base of the huge rocket as great clouds of steam shot out on either side of the water-cooled blast pit.

Slowly – almost too slowly – the vehicle began to rise on a pillar of flame. It was an incredible moment, as though the astronauts were being lifted by the sheer force of human desire. As the rocket began to clear the tower, hardened newsmen were on their feet shouting – GO! GO!

As the incredible noise of the engines crossed the miles which separated the onlookers from the spectacle – and the all-consuming reality of the moment began to leave its mark – people stood with tears streaming from their eyes as they gazed upward at the huge rocket roaring its defiance to the skies.

Tilting away from the Cape and climbing ever higher its exhaust crackled like a machine gun, the shock diamonds being plainly visible. Watchers on the ground saw clearly the first stage separate as stage two took up propulsion. Eleven minutes later the calm voice of Houston Mission Control had confirmed that the third stage had completed its initial burn and the astronauts were in Earth-orbit. Within 3 hr. the third-stage engine would re-ignite and put them on course for the Moon.

The names of the men who rode that ship are now legendary. The commander Neil Alden Armstrong, then thirty-eight, was a former civilian test pilot born in Wapakoneta, Ohio. The command module pilot was thirty-eight-year-old Michael Collins, then an Air Force Lieutenant-Colonel born in Rome, Italy. Lunar module pilot was Edwin Eugene Aldrin, Jr, thirty-nine, a Colonel in the US Air Force born in Montclair, New Jersey.

. . . . . .

Three days after setting out Apollo 11 was orbiting the Moon, with Armstrong and Aldrin preparing to separate their 'Eagle' landing craft from the 'Columbia' CSM.

After careful checks had been made in lunar orbit came the moment of truth. Collins was alone in the command ship and 'Eagle', on its own, was coming around the far side of the Moon to start the retro-burn which would commit the astronauts to a landing in the Sea of Tranquillity.

The procedure of the lunar landing is described and illustrated on pages 194–6. Reproduced below is the actual dialogue that passed between 'Eagle' and Houston Mission Control during the critical moments before and after touchdown:

*CAPCOM 60 seconds.

EAGLE Lights on. Down 2-1/2. Forward. Forward. Good. 40 feet, down 2-1/2. Picking up dust. 30 feet. 2-1/2 down. Faint shadow. 4 forward. 4 forward, drifting to the right a little. 6 (garbled) down a half.

CAPCOM 30 seconds.

EAGLE (garbled) forward. Drifting right. (garbled) Contact light. Okay, engine stop. ACA out of detent. Modes control both auto, descent engine command override off. Engine arm, off. 413 is in.

CAPCOM We copy you down, Eagle.

EAGLE (Armstrong) Houston, Tranquillity base here. The Eagle has landed.

CAPCOM Roger, Tranquillity, we copy you on the ground. You've got a bunch of guys about to turn blue. We're breathing again. Thanks a lot!

'Eagle' had touched down at 1618 hr. EDT on 20 July after the crew had negotiated craters and a boulder field; and commander Neil Armstrong stepped on to the lunar surface at 2256 hr. that evening followed by Aldrin.

In the strange low-gravity environment the astronauts began by inspecting the lunar module and testing the consistency of the soil which they described as powdery but cohesive; but their first uncertain steps had soon turned into exuberant kangaroo hops as they set about planting the Stars and Stripes in the moonsoil, laying out their scientific instruments, and collecting the first specimens of soil and rock. The total time of their EVA – from hatch open to hatch close – was 2 hr. 31 min. 40 sec.

* CAPCOM, Capsule Communicator, the single individual on Earth who talks directly with the crew.

Armstrong and Aldrin were on the Moon for 21 hr. 36 min. 21 sec. They lifted off at 1354 hr. EDT on 21 July bringing with them some 20·8 kg. (46 lb) of samples.

After re-joining Collins in the Moon-orbiting 'Columbia', the trio made their critical burn out of lunar orbit setting course for home. At 1250.35 hr. EDT on 24 July their capsule splashed down in the Pacific Ocean and President Nixon, who had talked by phone to Armstrong and Aldrin on the Moon, was aboard the recovery ship USS *Hornet* to greet them.

Four months after this heroic exploit, Apollo 12 was heading astronauts Captain Charles Conrad, Jr, and Captain Alan L. Bean for the Ocean of Storms; the command pilot was Richard F. Gordon, Jr.

The launching, which came on 14 November under conditions of rain and cloud, was not without incident. Barely 50 sec. after lifting off two lightning flashes were seen from the ground and almost at the same time Conrad was heard to say: 'Just lost the platform. . . . Everything in the world just dropped out.'

Having arrived in lunar orbit Conrad – still awed by the experience – was confirming what Mission Control already suspected – that the rocket had been struck by lightning. There were scorch marks on the side of the service module.

Although Apollo 12 survived the experience, NASA immediately revised launch weather regulations so that no such incident should happen again.

Not only did the landing by the lunar module 'Intrepid' in the Ocean of Storms achieve pin-point accuracy, being within walking distance of the Surveyor 3 moonprobe which had arrived two and a quarter years before, but the men were able to emplace the first nuclear-powered Apollo Lunar Surface Experiments Package (ALSEP) on the lunar surface for continuous scientific reporting. They also were able to make two excursions on foot outside the lunar module lasting 3 hr. 56 min. and 3 hr. 49 min. respectively, obtaining 34·1 kg. (75 lb) of lunar samples.

The homeward flight passed without major incident and on 24 November the command module landed in the Pacific after a flight which had lasted just over 10 days $4\frac{1}{2}$ hr.

The huge success of these two missions meant that John Kennedy's call for a man on the Moon during the decade of the sixties had been exceeded by a handsome margin. *Four* men had walked on the Moon – and still there was no response from the Soviet Union.

However, when Apollo 13 went to the launch pad it was not just the superstitious who felt apprehensive. At this scale of complexity could the high skills of Apollo – and 'lady luck' – continue to hold?

Apollo 13 roared into the blue on 11 April 1970 with the intent of landing astronauts James A. Lovell, Jr, and Fred W. Haise in a hazardous upland area of Fra Mauro. Command module pilot was John L. Swigert, who at a late stage had replaced Tom Mattingly because of worry that he might have contracted German measles.

A cause for concern during the launching was the premature shutdown (2 min. 12 sec. early) of the centre engine in the S-II second-stage J-2 cluster which necessitated burning the four outer engines for a correspondingly longer period and running the single J-2 engine of the third stage longer than planned. That problem overcome the spacecraft was successfully injected into the desired lunar trajectory.

Then, two days later at 2211 hr. EST – with the astronauts some 329,845 km. (205,000 miles) from Earth – instruments at Houston Mission Control registered a major malfunction and the voice of astronaut Lovell was heard reporting: 'We have a problem!'

At first it seemed that the ship had been struck by a meteoroid. In fact, an oxygen tank of the service module's fuel-cells had exploded causing an immediate loss of electrical power.

The crew were now face to face with the problem of survival. With a Moon landing out of the question, technical staffs at NASA and the spacecraft contractors were summoned from their homes to work out rescue procedures. An entirely new set of flight instructions had to be computed at Houston and passed up to the Apollo 13 crew.

As the craft was already committed to the Moon, there was no possibility of immediately recalling the mission and the astronauts had to wait for their craft to swing around the Moon and return. In agreement with Mission Control it was decided to retain the lunar module for as long as possible in order to utilize oxygen and electrical supplies in that vehicle. For most of the remainder of the flight the men employed the lunar module as living quarters, using the LM descent engine to make necessary trajectory corrections.

Finally, approaching the Earth on 17 April, the service module was jettisoned. It was only then that the crew saw the full extent of the damage. The explosion had exposed a large section of the equipment bay; the 3·96 m. (13 ft) by 1·82 m. (6 ft) access door was entirely missing and the inside had been severely damaged.

The crew, having transferred back into the 'Odyssey' command module and sealed the hatch, released the lunar module which had served them so well as a 'life-boat'; and to everyone's relief their capsule made a perfect splash-down in the Pacific after an agonizing 142 hr. 54 min. 41 sec. of flight.

Two months later the Apollo 13 Review Board announced its findings. A short-circuit had ignited insulation in the spacecraft oxygen tank no. 2, causing failure of the tank. The Board recommended the command and service module systems be modified to eliminate potential combustion hazards in high-pressure oxygen of the type revealed by the accident (see page 280).

All subsequent spacecraft were modified in accordance with these regulations and Apollo 14 was launched with the same lunar objective – Fra Mauro – at 1603 hr. on 31 January 1971.

The mission, flown by Alan B. Shepard, Jr, and Edgar D. Mitchell, with Stuart A. Roosa as command module pilot, was an enormous success despite initial docking difficulties between the CSM 'Kitty Hawk' and LM 'Antares' soon after the spacecraft had been inserted into the trans-lunar trajectory. Docking succeeded at the *sixth* attempt.

For the first time a new approach technique was adopted for the lunar landing. The complete spacecraft – CSM and LM – was made to dip within 17 km. (10·5 miles) of the Moon's surface before separating, a manœuvre which gave the LM an extra

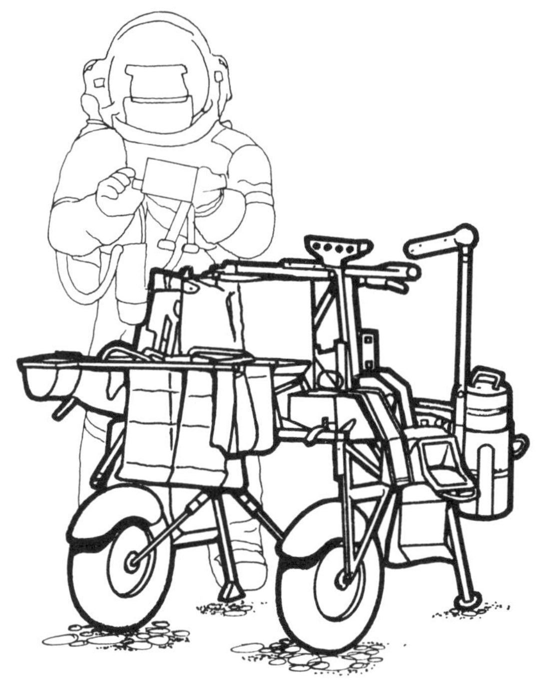

Mobile Equipment Transporter (MET) used by the Apollo 14 astronauts to carry tools and rock samples.

15 sec. of hover time in the final approach phase of the landing.

After some anxiety due to a fault in the LM computer the astronauts touched down on the lunar surface at 0417 hr. EST on 5 February within 18·3 m. (60 ft) of the intended target; and during two EVA periods of 4 hr. 50 min. and 4 hr. 35 min., Shepard and Mitchell deployed and activated another ALSEP including an array of geophysical instruments which transmitted data on the Moon's interior and exterior environment to Earth. The astronauts took with them a two-wheeled handcart – the Mobile Equipment Transporter (MET) – to help with the transportation of tools and rock samples. Before boarding the LM for the last time Shepard played his famous golf shot, landing the ball in a crater about 18·3 m. (60 ft) away.

Some 42·6 kg. (94 lb) of lunar samples were collected during this mission, which included two rocks weighing about 4·5 kg. (10 lb) each. After spending 33½ hr. on the Moon, the lunar module ascent stage lifted off from the surface at 1347 hr. on 6 February 1971.

The splash-down was made in the South Pacific at 1605 hr. EST three days later.

Apollo 15, launched at 0934 hr. EDT on 26 July 1971 headed for another highland region of the Moon – the Hadley Apennine. This time lunar transportation entered the executive class with the inclusion of an electric-powered 'moon-jeep', or Lunar Roving Vehicle (LRV), which the astronauts unpacked from its stowage in the descent stage of the lunar module after landing.

The team comprised David R. Scott (commander), command module pilot Alfred M. Worden, and lunar module pilot James B.

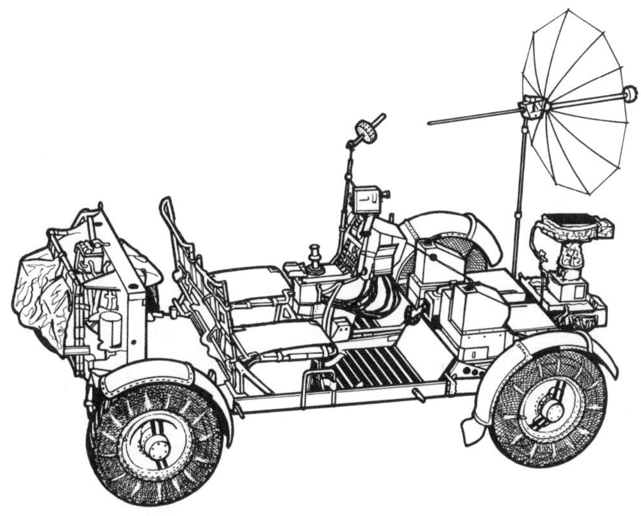

Electric-powered Lunar Roving Vehicle (LRV) which the Apollo 15, 16, and 17 astronauts used to enlarge their radius of exploration. Colour TV cameras with remote azimuth, elevation and zoom controlled from Houston Mission Control operated while the LRV was parked.

Irwin. Modifications to the spacecraft (page 282) permitted a longer stay on the lunar surface and additional experiments in lunar orbit.

Scott and Irwin landed their 'Falcon' lunar module on target at Hadley Apennine at 1816 hr. EDT on 30 July. During their 66 hr. 55 min. stay they made three EVAs lasting a total of 18 hr. 36 min., deploying an ALSEP with geophysical instruments and making extensive forays in the LRV to explore geological features. Scott demonstrated with the aid of a geological hammer and a falcon's feather that objects fall at the same rate in a vacuum. The astronauts collected some 76·6 kg. (169 lb) of lunar samples from a wide range of sites.

In the 'Endeavour' command module Alfred Worden conducted scientific observations of his own account, operating two cameras and gamma ray and X-ray sensors mounted in the service module.

After the surface explorers had rejoined Worden in lunar orbit, a sub-satellite was released and after making 74 lunar revolutions the craft began its earthward journey. Astronaut Worden made an EVA to retrieve film from the lunar cameras in the SM during the early part of the trans-Earth coast period.

The command module splashed down in the Pacific Ocean at 1646 hr. EDT on 7 August 1971 after some concern for the landing when one of the three main parachutes failed to inflate properly. The failure was later to be attributed to contamination during the dumping of fuel from the CM's attitude control system.

Apollo 16, the fifth lunar landing mission, was launched at 1254 hr. EST on 16 April 1972 with astronauts John W. Young, Thomas K. Mattingly, II, and Charles M. Duke, Jr.

The lunar module 'Orion' touched down in the Descartes highlands as planned on 20 April at 2123 hr. EST. Young and Duke set records for the longest stay on the Moon, 71 hr. 2 min., and the longest lunar surface EVA time totalling 20 hr. 14 min. They set out another ALSEP, operated the first 'astronomical observatory' on a celestial body – a Far UV Camera/spectrograph, and collected some 95·2 kg. (210 lb) of lunar samples.

Astronaut Mattingly spent 3 days 9 hr. 39 min. orbiting the

Moon in the 'Casper' CSM during the absence of his colleagues, operating a complex array of scientific instruments including two lunar mapping cameras and observing geological features.

Before departing lunar orbit a sub-satellite was released from the command ship and on the earthbound trip Mattingly spacewalked for 1 hr. 24 min. to retrieve film canisters from the lunar cameras in the service module.

The spacecraft splashed down in the Pacific Ocean at 1445 hr. EST on 27 April.

On 23 May three mortars set up on the Moon by the Apollo 16 astronauts were fired by remote control from Earth over distances of 152 m. (500 ft), 305 m. (1,000 ft) and 915 m. (3,000 ft) from seismometers in the ALSEP station and the results relayed to Earth.

The sixth and last Apollo mission to the lunar surface – the first night launching – began at 0033 hr. EST on 7 December 1972. Led by Eugene Cernan with Ronald Evans as CSM pilot, the astronaut team was completed by LM pilot Harrison 'Jack' Schmitt, a qualified geologist.

From roadside parking lots near the Kennedy Space Center and along Cocoa Beach, thousands of people saw the immense torch of Saturn 5 light the launch area as bright as day in the most spectacular lift-off in rocket history.

It was a tough assignment, the aim being to set down the 'Challenger' lunar module at Taurus Littrow, a lunar valley which nestles between majestic mountains on the south-eastern rim of the Sea of Serenity. This was the most easterly site selected for Apollo exploration.

Cernan and Schmitt landed at 1355 hr. EST on 11 December to begin the most extensive sortie of the entire lunar programme. The record speaks for itself: longest surface stay-time, 74 hr. 58 min. 38 sec.; longest total surface EVA time, 22 hr. 5 min. 6 sec.; longest single surface EVA time, 7 hr. 37 min. 21 sec.; longest lunar distance travelled with the Lunar Roving Vehicle on one EVA, 19 km. (12 n. miles), and on three EVAs 35 km. (22 n. miles).

The final EVA ended with the unveiling of a plaque on the descent stage of 'Challenger' signed by the three astronauts and

President Nixon, commemorating the end of Apollo exploration. President Nixon sent this message: 'This may be the last time in this century that men will walk on the Moon. But space exploration will continue, the benefits of space exploration will continue, the search for knowledge through the exploration of space will continue, and there will be new dreams to pursue based on what we have learned.'

Lifting off from the Moon at 1755 hr. EST on 14 December, the exploration team was soon back aboard the orbiting CSM as the world looked on by TV. During the return to Earth Evans made a 1 hr. 7 min. 18 sec. space-walk to retrieve film and other experiment data from the service module.

At 1424 hr. EST on 17 December, the Apollo 17 command module 'America' splashed down in the Pacific. It had been the longest Apollo flight lasting 12 days 14 hr. and returning the most lunar samples, 116·5 kg. (257 lb).

. . . . . .

Thus ended Man's first epic series of Moon explorations – a lasting memorial to the skill, courage, and political will of the American nation. More than 363 kg. (800 lb) of rock and soil samples, from lunar 'seas' and highlands, had been brought to Earth and a network of nuclear-powered automatic research stations left to work unattended at five different landing sites. These stations continued to return data on lunar impacts, sub-surface disturbances ('moonquakes'), temperatures, etc., allowing the research programme to continue for years after the astronauts had returned.

Examination of recovered moon rocks left no doubt of the Moon's volcanic history; possibly there had been at least four separate melting periods. Many of the samples were similar to terrestrial volcanic rocks although there were considerable differences in chemical make-up, with more of the higher melting-point elements such as chronium, titanium, yttrium, and zir-conium.

A report to the third Lunar Science Conference emphasized the three basic materials present on the Moon: *basaltic rock*, rich in iron, and differing from the common basalt formed on Earth; *kreep*, a material, rich in potassium and high in radioactive elements, and *plagioclase*, which is low in radioactivity and high in aluminium, calcium, and sodium.

A so-called 'Genesis' rock found by the Apollo 15 astronauts was provisionally dated at the State University of New York at about 4,150 million years. The key to the rock's age was the relationship between potassium and argon.

However, an even older specimen was found by the Apollo 16 team which the same University group dated at about 4,250 million years. One of a number of rock samples taken from the rim of the North Ray Crater, it was a lightweight *breccia* – compressed dust and fragments solidified by heat and containing crystals of plagioclase, feldspar, and pyroxene.

It was during the Apollo 17 excursion that the astronauts found at the rim of Shorty Crater – named after Shorty Powers of 'Voice of Project Mercury' – orange-red soil which at first led to speculation that it might have been formed by volcanic action. In fact it appears to have resulted from a meteoroid impact. When examined under a microscope the soil appeared in the form of minute orange glass beads.

Dr Edwin Roedder of the US Geological Survey said the glass was completely free of bubbles, and many beads were broken. The deep orange colour, he supposed, stemmed from the high titanium and iron contents. Some of the beads were crystallized, and hence were black.

Concerning the Moon's magnetic field information sent by the ALSEP stations showed it to be about 1,000 times weaker than the Earth's. Although the situation could have been very different in a past epoch, there was no significant dipole field today and a magnetic compass on the Moon would be useless.

This contrasts with the Earth's magnetic field which is generated deep within the planet by the constantly rotating molten metal core. The core develops a field measurable many thousands of miles out in space.

The main lunar magnetic field, on the other hand, consists of near-surface fields highly variable in magnitude and direction.

From magnetic data Dr Palmer Dyal, a NASA lunar investigator at Ames Research Center, California, calculated the abundance of free iron on the Moon at about 2·5 per cent by weight. Total iron content of the Moon is about 9 per cent by weight, which compares with about 30 per cent for the Earth.

Although seismic activity on the Moon was on a far smaller scale than on Earth the ALSEP stations confirmed the existence of 'moonquakes' associated with activity deep within the Moon at 676–1,116 km. (420–720 miles), and with shallow activity produced by thermal heating and cooling during the day and night. There was no evidence of mountain building or continental drift.

When discarded rocket stages and LM ascent stages were sent crashing into the Moon, seismic signals often reverberated for an hour; on Earth they would have been damped out within minutes.

The difference was explained by the diffusive propagation of the shock waves as a result of intense scattering, particularly near the lunar surface. According to NASA, the effect was enhanced by low attenuation due to the lack of water and other volatiles in the pores of the lunar rocks.

The seismic data indicated that the lithosphere (solid part) of the Moon is 676–1,116 km. (420–720 miles) thick, much thicker than our own planet. The Moon's core is probably near the melting point.

Other findings suggest that, in comparison to the Earth, the Moon is depleted in volatile elements like iron, sodium, and potassium.

From the evidence of Apollo it seemed unlikely that the Moon was once part of the Earth. More likely the two bodies were formed together, or the Moon is a captured body.

Fallen Astronauts

'Fallen Astronauts' are commemorated by a tiny aluminium figure which the Apollo 15 explorers left in the moondust at Hadley Base. Beside it lies a plaque on which is inscribed 14 names:

Virgil Grissom, Edward White, and Roger Chaffee, who perished in the Apollo fire (27 January 1967); Theodore Freeman, whose training aircraft collided with a goose while coming in to land (31 October 1964); Elliot See and Charles Bassett, whose plane struck a rooftop at the McDonnell factory while landing in bad weather (28 February 1966); Edward Givens, who died near Houston, Texas, when his car hit an embankment and overturned (6 June 1967); Clifton Williams, whose plane flew into the ground during a training exercise (5 October 1967).

Soviet cosmonauts commemorated are: Vladimir Komarov, who plunged to his death in the Soyuz 1 capsule when parachute lines tangled (24 April 1967); Yuri Gagarin* – first man in space – whose two-seat MiG-15 crashed during a training flight (27 March 1968), and Pavel Belyaev, who died from complications following an operation for stomach ulcers (10 January 1970). Three more cosmonauts lost their lives when their Soyuz 11 lost pressure while returning from orbit (30 June 1971). They were Georgi Dobrovolsky, Viktor Patseyev, and Vladislav Volkov.

* With Gagarin died Colonel-Engineer Vladimir Seryogin, a non-cosmonaut.

CHAPTER FIVE

SKYLAB

No finer demonstration of the value of man-in-space could possibly have been contrived than the events which followed the launching on 14 May 1973 of Skylab, America's first experimental space-station. Years of concentrated effort and a financial investment of around £1,000 million seemed to have gone for nought 63 sec. after lift-off when the meteoroid shield was torn away from the exterior of the cylindrical workshop, carrying with it one of the station's retracted solar 'wings'. Even the second solar array (as it subsequently transpired) had not opened and appeared to have suffered damage. As a result the overheated and underpowered Skylab seemed to be stranded beyond recall.

The remarkable salvage operation devised by the NASA/industrial team began on 25 May when Charles Conrad, Dr Joseph Kerwin, and Paul Weitz ascended in their Apollo CSM from the Kennedy Space Center at 1400 hr. BST carrying with them three types of hastily improvised tools and sun shields.

After achieving rendezvous with Skylab some 7½ hr. later, a fly-around inspection showed the extent of the damage. Not only had one solar panel been completely torn away but the other was held barely open by a strip of torn metal from the meteoroid shield. The loss of the shield meant that the orbital workshop was exposed to the full impact of solar heating and there was a danger that internal temperatures would reach levels which could seriously degrade systems and stored food.

Standing in the Apollo hatchway the spacesuited Weitz spent an hour trying to release the remaining solar panel by prodding at it with a rod, but to no avail.

More troubles were in store when repeated attempts to dock with the station were unsuccessful. In the end the astronauts had to put on their spacesuits again, remove the drogue from the docking tunnel, and attach a by-pass lead to switches of the docking mechanism.

After sleeping in the Apollo cabin, the men finally entered the overheated Skylab on 26 May without spacesuits, but wearing face masks in case toxic gases had leaked from the station's internal insulation due to the high temperature conditions. Although metal surfaces were hot to the touch, internal conditions were much better than expected and the team lost no time in deploying a 6·7 m. × 7·3 m. (22 ft × 24 ft) parasol-type heat shield from a scientific airlock in the side of the workshop. The parasol was made of layers of nylon, Mylar, and thin aluminium foil.

Skylab 4 On-board photograph. An overhead view of the Skylab space-station in Earth-orbit as photographed from the Skylab 4 command module during the final fly-around by the CSM. The space-station is contrasted against a cloud-covered Earth. One of the Orbital Workshop solar array system wings is missing. A solar shield was deployed to cover a portion of the OWS where a micrometeoroid shield was torn off.

This produced a rapid drop in temperature, and by the fourth day the astronauts had settled down to some kind of routine.

During this early period electrical supplies came from the cruciform solar panels of the Apollo Telescope Mount (ATM) which had deployed correctly in orbit on launch day and were undamaged.

It was therefore possible to begin certain experiments. On 29 May the astronauts used the ATM for the first time to obtain solar spectra and the following day the first Earth-resource photographs were taken.

In the meantime, preparations were being made in consultation with Mission Control to release the jammed solar 'wing'. On 7 June the spacesuited Conrad and Kerwin emerged from the Skylab airlock module guided by Schweickart at Houston who had practised various procedures in the trainers at the Marshall Space Flight Center in Huntsville. They cut through the metal strip holding the solar panel and after some difficulty got the 'wing' open to its fullest extent.

Meanwhile, spaceflight records were already beginning to tumble. On 4 June Conrad had exceeded James Lovell's 715 hr. 5 min. world record for the longest time spent by a man in space in multiple missions. Two days later Skylab itself had exceeded the US duration record for a single manned spaceflight set by the Gemini 7 crew at 330 hr. 35 min.

At 0822 hr. BST on 18 June the trio set a new absolute record, passing the 23 days 18 hr. 21 min. set by the ill-fated crew of Soyuz 11.

The next day Conrad and Weitz space-walked from Skylab for 96 min. to replace ATM film canisters and repair one of 18 storage batteries (by tapping a regulator with a small hammer) restoring 200 W of power.

Having logged the full period assigned to them, the astronauts undocked from the station at 0858 hr. GMT on 22 June; they splashed down in the Pacific at 1350 hr. GMT about 1,335 km. (830 miles) south-west of San Diego. On this occasion there was no attempt to remove the astronauts from the capsule by helicopter but instead they were lifted on to the deck of the recovery carrier, USS *Ticonderoga*, and transferred below decks still in the capsule.

Although further repairs to Skylab were needed, the major part of the salvage operation had been accomplished and the first experimental results returned to Earth. Apart from the ATM experiments which were beginning to reveal new depths of the Sun, multispectral photographs of the Earth had included strip mining in Kentucky, oil fields in Venezuela, and ocean currents in the Atlantic.

Particularly rewarding was the apparent ease with which the first boarding party had stood up to the uncertainties surrounding a month-long orbital flight, despite the considerable demands placed upon them by the salvage operation.

To ensure the maximum return from the ailing Skylab, the launching of the second crew was brought forward. At 1208 hr. BST on 28 July Commander Alan Bean, Dr Owen Garriott, and Major Jack Lousma took off in pursuit of the station, docking with it 7½ hr. later. After transferring into Skylab with a small menagerie of biological test subjects – including mice, fish, gnats, and two spiders – the crew experienced symptoms of motion sickness which in the case of Lousma were quite severe. To add to the crew's difficulties a leaking hatch, noticed by Mission Control had to be rectified which entailed waking the men; and although a first full day of experiments was carried out on 1 August, a planned space-walk to fix a new sunshade had to be deferred.

Then came the more serious report that two of four sets of attitude control thrusters on the Apollo ferry were leaking fuel, reducing the craft to 'minimum flyable condition' and hazarding the astronauts' safe return. Immediately, NASA put into effect contingency plans: either the crew would be ordered to return before the situation deteriorated still further, or they would be asked to continue in orbit until a five-seat rescue craft could be made ready.

NASA chose the latter expedient and work began on converting the third Apollo ferry for a possible launch on 5 September. If needed, this would be crewed by two men – ex-test pilot Vance Brand and Dr Don Lind. In the meantime, the Skylab team were asked to defer their space-walk to fix the new sunshade until 6 August.

As it turned out the problem was not quite as serious as it might have been. The fault was traced to a leaking valve seating and a faulty pipe.

With confidence returning for a successful mission, two of the Skylab astronauts made a record space-walk of 6 hr. 31 min. to replace the original 'parasol' with a larger and more efficient sunshade which quickly dropped the temperatures inside the workshop by 5°, to an average of about 75° F. During this EVA the astronauts also replaced film canisters in the ATM, inspected damage to the Apollo CSM attitude control system and found the cause of a leak which had developed in the cooling system of the workshop.

On 13 August Bean successfully tested the 102 kg. (225 lb) backpack Astronaut Manœuvring Unit (AMU) within the capacious interior of the workshop working 14 nitrogen gas directional thrusters with two hand controllers. (A similar device was to have been tested by Eugene Cernan during the Gemini 9 mission in 1966 – page 271.)

Of the other experiments the mice and gnats had died due to an electrical fault in their containers; but both the spiders (one of a number of student experiments) spun normal webs, and the fish learnt to swim, despite their weightless condition.

However, on 16 September, Anita, one of the spiders, was found dead in its cage, the probable cause, according to Dr Garriott, being a diet of filet mignon instead of live flies! (Arabella, the second spider, died within three days of being returned to Earth.)

As the mission wore on the astronauts had lost all symptoms of motion sickness and made up for lost time by carrying out a full range of medical, physical, astronomical, and Earth-resource experiments. Important results were obtained by melting and forming metals in a small electric furnace. Large crystals were also grown under weightless conditions.

On 22 September Bean and Garriott space-walked to replace film canisters in the ATM ready for the third crew visit and began preparations for stowing film and taped records and other experimental results aboard the Apollo ferry. As the rescue operation was no longer deemed necessary, the return was made in the original Apollo CSM despite the thruster problem.

To everyone's relief, once again NASA's technical judgement triumphed. The command module splashed down in the Pacific at 1120 hr. BST on 25 September after a record spaceflight of more than 40 million km. (24 million miles) lasting 59 days 1 hr. 9 min.

The capsule was hoisted on to the deck of the helicopter carrier USS *New Orleans* with the astronauts aboard. The scientific return included 17,000 photographs of the Earth, 75,000 images of the Sun, and 40 million km. (25 million miles) of taped data. If anything, the men were in better physical shape than their predecessors.

The third and final boarding party – Lieutenant-Colonel Gerald P. Carr, Dr Edward G. Gibson, and Lieutenant-Colonel William R. Pogue – lifted off on 16 November after nine days' delay caused by the discovery of hairline cracks in the stabilizing fins of the Saturn 1B launch vehicle. This had necessitated replacing the fins.

Eight hours after leaving Cape Canaveral the team had docked with Skylab at the third attempt. After settling in the full programme of experiments was resumed and on 22 November Pogue and Gibson space-walked for a record 6 hr. 34 min. 35 sec. to repair an antenna and replace film in the ATM.

The next day the first photographs were taken of the comet Kohoutek, and then came word of another major setback. One of the three Control Moment Gyroscopes (CMGs) which adjusted Skylab's attitude had to be stopped because the bearings had overheated.

Skylab had three of these 55·9 cm. (22 in.) 110 kg. (242 lb) inertial wheels encased within separate 99 cm. (39 in.) spheres mounted on three sides of the ATM. Spinning at 9,000 r.p.m. the electrically powered wheels had the task of controlling the space-station on three axes as an alternative to gas-jets.

Failure of one CMG meant making greater use of the gas-jet

attitude control system at the back of the workshop – normally used for large attitude changes – consuming precious nitrogen fuel. Whilst it was possible to work two gyros together, it would be impossible to continue if two gyros failed.

The astronauts nursed the station throughout the remainder of the mission despite concern for a second gyro which had dropped spin rate by 10–14 r.p.m. but which continued to operate, allowing further photography of Kohoutek.

On Christmas Day 1973 Carr and Pogue space-walked for a record 7 hr. to obtain more photographs of the comet, but attitude control manœuvres necessitated by the faulty gyros consumed 10 per cent of the remaining thruster fuel. Four days later Gibson and Carr space-walked for 3 hr. to photograph the comet which by then had developed a gas and dust tail with a gas spike, or anti-tail, caused by gas ejecta pointed at the Sun.

After a record 84 days in space the third Skylab team separated their Apollo ferry at 1034 hr. GMT and made a final fly-around inspection. They splashed down in the Pacific 281 km. (175 miles) south-west of San Diego at 1517 hr. GMT.

It was the final answer to the pessimists who had forecast that Man could not survive long exposure to weightless conditions and re-adaptation to normal gravity. Nine astronauts backed by the immense skill of the NASA/industrial team had brought another goal of the founders of Astronautics to fruition – the sequential manning of a large space-station in Earth-orbit.

The three Skylab crews spent 740 hr. observing the Sun with telescopes and brought home more than 175,000 solar pictures. These were the first consistent recordings of solar activity above Earth's obscuring atmosphere and the data began quickly to revolutionize long-standing theories of solar physics and may lead to the Sun's vast energy being used for practical purposes on Earth (see the companion volume *Frontiers of Space*).

The 46,000 photographs and 64 km. (40 miles) of electronic data tape obtained by Skylab's Earth-resources instruments and cameras were immediately put to daily use by government and industry for studies ranging from agriculture to zoology. Data obtained by the third Skylab crew on comet Kohoutek, coordinated

SKYLAB DIGEST

Summary of flights made by astronauts to the Skylab 1 space-station launched on 14 May 1973.

Mission	Crew	Notes
Skylab 2 25 May–22 June 1973	Charles Conrad, Jr, Joseph P. Kerwin, and Paul J. Weitz	Docked with Skylab 1 for nearly 28 days. Repaired damaged station; erected sunshade. Data obtained on 46 of 55 experiments; 3 space-walks totalling 5 hr. 41 min. Mission lasted 672 hr. 50 min.
Skylab 3 28 July–25 Sept. 1973	Alan L. Bean, Jack R. Lousma, and Owen K. Garriott	Docked with Skylab 1 for about 59 days. Made further repairs; erected larger sunshade. Replaced rate gyros; exceeded pre-mission plans for scientific experiments; 3 space-walks totalling 13 hr. 44 min. Mission lasted 1,427 hr. 9 min.
Skylab 4 16 Nov. 1973–8 Feb. 1974	Gerald P. Carr, Edward G. Gibson, and William R. Pogue	Docked with Skylab 1 for nearly 84 days completing experimental programme. Replenished coolant supplies, repaired antenna, observed comet Kohoutek; 4 space-walks totalling 22 hr. 21 min. Mission lasted record 2,017 hr. 16 min.

with world-wide observations from Earth, provided the greatest advance in cometary research since the studies of the British astronomer, Edmund Halley, in the late seventeenth and eighteenth centuries.

Skylab by any reckoning was a monumental triumph.

CHAPTER SIX

SALYUT

Russia's space-station programme evolved from experiments carried out with Soyuz spacecraft. Not only was it possible to make the necessary rendezvous and docking trials, essential to space station re-supply, but it was possible to conduct a number of preliminary scientific experiments.

One of Russia's most remarkable early achievements was the docking together in 1967 of two unmanned Soyuz-type spacecraft – Cosmos 186 and 188 – which introduced the opportunity to keep manned space-stations re-supplied by purely automatic means.

While the early Soyuz flights were in progress, Western intelligence sources were anticipating major developments bound up with the testing of the Soviet super-booster known in the West as the type G-1 (see the companion volume *Missiles and Rockets*) which had suffered a major setback at the Tyuratam cosmodrome. The large rocket was looked upon as the means of placing very large payloads into Earth-orbit, including the central core of a multi-man space-station.

As it turned out, however, there was to be an interim development. The first outward signs had already appeared in November 1969 when a large payload, apparently launched by a Proton-type D-1-H rocket, failed to achieve orbit. A similar more successful experiment in December 1970 was concealed under the label Cosmos 382.

These are believed to have been tests of structural prototypes of the Salyut space-station launched on 19 April 1971, which led to the first visit by a space crew to an orbiting space laboratory.

Lifting off from Tyuratam on 23 April Soyuz 10 took cosmonauts Colonel Vladimir Shatalov, Aleksei Yeliseyev, and Nikolai Rukavishnikov into an orbit of 208–249 km. (129–155 miles) inclined at 51·6° to the Equator. Then, in conjunction with Soviet ground stations and tracking ships linked by the Molniya 1A satellite, the craft was manœuvred towards a rendezvous with Salyut 1. Having completed 22 orbits (and the station 86) Soyuz 10 began final approach manœuvres under automatic control switching to manual when the distance between ferry and station was down to 180 m. (590 ft).

During the operation Soyuz made three orbit changes and Salyut three. Docking was achieved but to the surprise of Western observers the crew did not enter the station. Instead, *Tass* explained that the test was restricted to tests of 'a new docking system'.

Soyuz 10, having separated after 5½ hr., was retro-fired at 0159 hr. (Moscow time) on 25 April making an uneventful re-entry and return. The command module soft-landed about 120 km. (74 miles) north-west of Karaganda 41 min. later.

Two days later a Western tracking station reported that the station had been moved into a higher orbit at about 250–270 km. (155–168 miles) to prevent orbital decay.

The station remained unmanned for one and a half months and then, at 0755 hr. (Moscow time) on 6 June 1971, Soyuz 11 was launched with cosmonauts Air Force Lieutenant-Colonel Georgi Dobrovolsky, commander; Viktor Patseyev, test-engineer, and Vladislav Volkov.

Once again the necessary orbital manœuvres were performed and the craft was manœuvred towards Salyut 1, first automatically and then manually this time from a distance of 100 m. (328 ft).

After docking had taken place at 1045 hr. (Moscow time) on 7 June and systems had been electrically and hydraulically interconnected, the cosmonauts opened the docking hatch and entered the station through the internal tunnel.

With this accomplished *Tass* issued a communiqué explaining that the Salyut/Soyuz combination weighed more than 25 tonnes and that the orbital module of Soyuz formed part of the station for purposes of sleep and rest. The docking unit of the ferry had a feature to achieve 'more rigid tightening' of the spacecraft after docking so that their butt ends came as close together as possible, providing a better hermetic seal. (Whether an imperfect seal was the reason why the Soyuz 10 cosmonauts did not enter the station remained in doubt.)

Stated mission objectives of Soyuz 11 were to: (a) check on-board systems; (b) test the station's orientation, navigation, and manœuvre control; (c) study Earth's geography and geology, atmosphere, snow, and ice cover; (d) study physical and chemical processes and phenomena in the atmosphere and outer space in different parts of the spectrum; and (e) continue bio-medical studies related to work aboard space-stations.

On 8 June the Salyut/Soyuz 11 combination was boosted into a slightly higher orbit of 240–265 km. (149–165 miles) by the comparatively low thrust correction engine in Salyut. Cosmonauts reported that the thrust was barely felt.

On board the station was an 'Orion' astrophysical observatory described as 'a complex and highly accurate optical-electronic system'. One telescope and spectrograph was outside the pressurized hull of the station and another was inside. On 18 June Patseyev made the first of a number of experiments pointing the instrument at a star in the constellation of *Centaurus*. He obtained spectrograms on film in the short-wave range which cannot be obtained from the ground.

The action of the telescope was automatic. The cosmonaut simply directed the telescope towards a selected star using a sighting tube. The telescope then locked on to the star while photographs were taken.

Three days later Patseyev simultaneously directed the two telescopes at the star *Alpha Lyrae* to obtain spectrograms in two different UV regions of the spectrum. It was said that the accurate and coordinated work of the station commander Georgi Dobrovolsky and flight engineer Vladislav Volkov contributed greatly to the success of this experiment.

Viktor Patseyev had also staked a claim to be regarded as the first 'space gardener'. He had special responsibility for tending the plants on Salyut 1 and after doing his physical exercises each evening, he watered his 'crops' which were lit by fluorescent lamps. They included plants of the genus *Crepis* (Hawk's beard), flax, and marrow-stem kale (which can be eaten as a salad).

The range of other experiments included extensive medical tests and exercises; study of the high-frequency electron resonance on the station's transmitting antenna (which results in a weakening of radio signals) and observations of the Earth's surface and weather phenomena.

The three men transferred back to the Soyuz 11 ferry with the flight log and other results of their experiments, and undocked from the space-station at 2128 hr. (Moscow time) on 29 June 1971.

Flight engineer Volkov had been carrying out an intensive radio exchange with Mission Control. He reported on completed operations and the functioning of the ship's systems. 'This is Yantar 1,*' the mission commander Dobrovolsky was reported as saying. 'Everything is satisfactory on board. Our condition is excellent. We're ready for landing.'

Retro-fire began at 0135 hr. (Moscow time) on 30 June. Communications on S-band appear to have been lost at about the time Soyuz's orbital module should have separated followed by the service module. At this point the craft was still well above the atmosphere and not yet subject to high-g forces due to atmospheric retardation.

Following re-entry the command module's drogue parachute opened automatically followed by the main parachute at about 7 km. (4·3 miles) altitude. The landing in Kazakhstan was on target and a helicopter of the Soviet recovery team was hovering near by as the command module's soft-landing rockets fired 1 m. from touchdown.

Then came the shock. Upon opening the hatch, *Tass* reported, 'the recovery group found the crew in their seats without any signs of life'. The mission – the longest ever – had lasted 570 hr. 22 min.

A day after the tragedy Salyut 1 was again manœuvred into a higher orbit to prevent orbital decay. It was unofficially reported in Moscow that at the time of the accident another Soyuz was being prepared at Tyuratam to launch three more cosmonauts to continue the research programme aboard the station.

Instead the ashes of three brave pioneers of spaceflight were buried in the Kremlin wall.

* Call-sign of Soyuz 11 commander.

What happened? A Soviet Government Commission looking into the accident on 11 July spoke of a 'rapid pressure drop occurring inside the descent vehicle' some 30 min. from touchdown, because of 'a loss of the ship's sealing'. There were no signs of structural failure. 'A technical analysis has made it possible to establish a number of possible causes of the seal failure, the study of which is continuing', the Report concluded.

It was not until mid-1973 that the Russians disclosed to American colleagues concerned with the Apollo-Soyuz Test Project (ASTP) what actually occurred. When the Soyuz orbital module separated from the re-entry module, shock waves produced by the explosive charges caused a pressure equalization valve in the command module to open prematurely releasing air from the cabin so that the men died from lack of pressure and oxygen.

In the meantime, on 11 October 1971, Salyut 1 had been deliberately destroyed by Soviet Mission Control when a retro-burn caused the station to re-enter the atmosphere and burn up harmlessly over the Pacific Ocean. It had circled the Earth for 175 days.

SOME UNMANNED COSMOS TEST-FLIGHTS THOUGHT TO BE RELATED TO THE SOVIET SOYUZ AND SALYUT PROGRAMMES.

Cosmos	Launch Date	Lifetime	Remarks
379	24 Nov. 1970	8 years	Manœuvre tests of Soyuz-type vehicle with up-rated service module propulsion (?).
382	2 Dec. 1970	1,000 years	Salyut manœuvre test (?).
398	26 Feb. 1971	20 years	As Cosmos 379 (?).
434	12 Aug. 1971	10 years	As Cosmos 379 (?).
496	26 June 1972	6 days	Test of re-designed Soyuz for two-man crew after Soyuz 11 depressurization accident.
557	11 May 1973	11·11 days	Salyut test
573	15 June 1973	2 days	Test of Soyuz ferry without solar cell arrays.
613	30 Nov. 1973	60 days	Soyuz 60-day test with systems inactive for most of mission in simulation of docking with Salyut.
638	3 Apr. 1974	9·9 days	ASTP test.
656	27 May 1974	2 days	Test of automatic Soyuz docking system (?) – see Soyuz 15.
672	12 Aug. 1974	5·9 days	ASTP test.

Testing of a modified Soyuz spacecraft following the Soyuz 11 tragedy began with the launching of an unmanned vehicle Cosmos 496 on 26 June 1972; the re-entry module was recovered after a flight lasting 6 days.

There were unconfirmed reports of a Salyut launch which went astray on 30 July 1972. If true, there is no evidence that it achieved orbit.

It was not until 3 April 1973 that another orbital station, Salyut 2, was ready on the Tyuratam launch pad. Launched at about 0900 hr. GMT, it entered an orbit ranging between 215 and 260 km. (134 and 161 miles) inclined at 51·6° to the Equator. *Tass* announced that the object was 'to perfect the design of on-board systems and equipment and conduct scientific and technical research in spaceflight'.

Within 24 hr. Soviet Mission Control had manœuvred the station to achieve a more stable orbit of 239–261 km. (148–162 miles). At the same time US tracking stations had detected more than 20 metallic fragments associated with the launching which seemed to point to the explosion of the separated carrier rocket.

On 8 April the orbit of the station was again changed – to 261–286 km. (162–178 miles). Three days later *Tass* reported that Salyut 2 had completed 130 revolutions 'with on-board systems functioning normally'. The orbit was 261–296 km. (162–184 miles).

Western observers waited for the expected launching in anticipation of an improved Soyuz ferry which would put cosmonauts aboard the station. But, on 18 April, unofficial sources in Moscow were quoted by UPI as saying that there were no plans for a manned flight to link with the space-station which (the report continued) was 'carrying out experiments connected with the joint US/Soviet flight planned for 1975'.

Speculation that Salyut 2 had been damaged during the launching grew and on 25 April tracking data from Goddard Space Flight Center indicated that the space-station had developed an uncontrolled rate of tumble and was breaking up. Three days later a final *Tass* communiqué stated that Salyut 2, 'having checked the design of improved on-board systems and carried out experiments in space, had completed its flight programme'.

In the summer of 1973 some light was thrown on what the Russians may have been attempting by the exhibition of a full-size engineering mock-up of a redesigned Salyut at the Star Town cosmonauts' training centre north of Moscow. Whereas Salyut 1 had fixed wing-like extensible solar panels fore and aft, the new design dispensed with this arrangement and instead employed a set of three pivoting solar panels attached to the station's 3 m. (9·8 ft) centre-section. Other discernible modifications included the rearrangement of pressure bottles for the life-support system and the introduction of an airlock hatch on the side of the docking module for extra-vehicular activities.

It seemed likely that Salyut 2 was one of these redesigned stations. Whether the shrapnel effect of a flying fragment from the exploded rocket stage had damaged the station, or the tumbling was caused by a 'stuck' attitude control jet the Russians did not say.

What looked like another attempt to launch a Salyut space-station came on 11 May 1973 when a large payload went into orbit from Tyuratam. According to data issued by the Royal Aircraft Establishment, Farnborough, the orbit ranged between 214 and 243 km. (133 and 151 miles) inclined at 51·59° to the Equator. The object, which had separated from its carrier rocket, was about 12 m. (39 ft) long by 4 m. (13·1 ft) diameter and probably had a mass of some 19,400 kg. (42,770 lb).

However, it was soon apparent that the vehicle had stuck in a low orbit because of control or engine problems; and after circling the Earth for 11 days it plunged back into the atmosphere and burned up.

In the meantime testing had begun of an improved version of the Soyuz spacecraft, at first unmanned within the Cosmos programme and then with two-man flight crews.

Soyuz 12 – the first manned launching after the depressurization tragedy of 30 June 1971 – ascended from Tyuratam at 1518 hr. (Moscow time) on 27 September 1973 with cosmonauts Vasily Lazarev and Oleg Makarov.

It was only now that Western observers became aware of the extent of modifications which had been made since the accident to improve the spacecraft's safety and reliability. As anticipated,

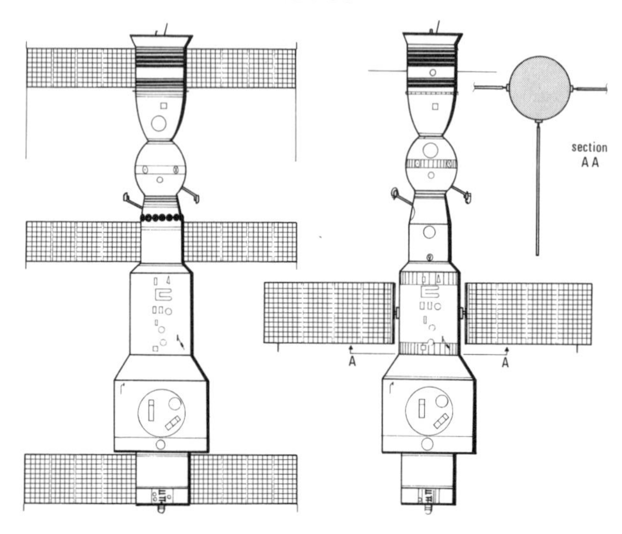

Soviet space-stations compared: *left,* Salyut 1; *right,* Salyut 3. In each case a Soyuz ferry is shown in the docked position. In the case of Salyut 3 the ferry dispensed with extensible solar wings and relied instead on internal batteries.

the third crew-member had been replaced by a life-support system allowing a two-man crew to wear lightweight spacesuits for added protection during launching, docking, and undocking, and the separation of spacecraft modules. The suits were removed during operations in orbit.

Externally, the most obvious difference was that Soyuz no longer had extensible solar wings but instead depended on internal chemical batteries for power supply. The Russians argued that since the space-station ferry required only about two days of independent flight operation, solar cells were unnecessary. After Soyuz had docked with Salyut, its batteries could be re-charged from the large steerable solar panels of the space-station itself.

Thus, there were now three versions of Soyuz:

1. An improved version of the original spacecraft used for

independent flight operations but crewed by two men instead of three.

2. The modified vehicle being developed for the 1975 joint US-Soviet Apollo-Soyuz Test Project (ASTP); and
3. The stripped-down vehicle (without solar 'wings') for space-station ferry missions.

The stated objectives of Soyuz 12 included checking and testing improved flight systems, further testing of manual and automatic controls in various flight conditions and spectography of separate regions of the Earth – from visible to infra red – to obtain data 'for the solution of economic problems'.

After circling the Earth in a preliminary orbit of 194–249 km. (120–155 miles) at 51·6° to the Equator, on 28 September the craft was manœuvred upward to achieve 326–345 km. (203–214 miles), contact with Soviet Mission Control being maintained via the tracking ship *Akademik Sergei Korolyov* in the Atlantic and the Molniya 1A communications satellite.

During the flight the cosmonauts made a special survey of the Earth using a camera that photographed the surface in nine different parts of the spectrum. Small biological experiments were also carried.

The Soyuz 12 capsule soft-landed some 400 km. (248 miles) south-west of Karaganda, Kazakhstan, at 1434 hr. (Moscow time) on 29 September 1973.

It was not long before the improved long-duration version of the Soyuz family was ready for test. On 18 December 1973 at 1455 hr. (Moscow time), Soyuz 13 roared from the launch platform at Tyuratam with cosmonauts Air Force Major Pyotr Klimuk and civilian flight engineer Valentin Lebedev. The craft, packed with experiments related to future space-station activity, had the following objectives: (a) observation of stars in the UV range using the Orion 2 telescope; (b) test of Oasis 2 closed-cycle biological experiment; (c) Earth-resource observations; (d) comprehensive verification and check-out of on-board systems; and (e) further testing of manned and autonomous control and methods of autonomous navigation under various flight conditions.

According to US tracking information, the craft began orbiting the Earth at a height of 188–246 km. (117–153 miles) inclined at 51·56° to the Equator. On the fifth revolution it was manœuvred upward to achieve 225–272 km. (140–169 miles).

Some of the experiments were an extension of those made by the ill-fated Soyuz 11 flight crew two and a half years before. Oasis 2 served to 'investigate the peculiarities of the growth and cultivation of individual biological subjects in weightless conditions'. The apparatus comprised two interconnecting cylinders one of which contained water-oxidizing bacteria which depended for their growth upon hydrogen obtained as a result of electrolysis. The other cylinder contained urobacteria 'which can break down compounds containing nitrogen and accumulate protein mass'. In this closed system, the waste products of the synthesis of one type of bacteria become the basic material enabling the other type to produce protein.

Apparently, the experiment was part of a Soviet plan to demonstrate the feasibility of biological closed-cycle life-support systems for large space-stations and other long-duration manned space systems which, eventually, could include interplanetary spaceships and extra-terrestrial bases.

The Orion 2 telescope system obtained spectrograms of more than 3,000 stars of different spectral classes in a band which can be examined only from space owing to absorption effects of the Earth's atmosphere. The chief designer of Orion 2, Grigor Gurzadyan, said 10,000 short-wave spectrograms obtained during the 8-day flight showed clearly stars of up to 10th magnitude and, in one case, in the region of sky around the star *Capella* in the constellation *Auriga*, stars of 11th magnitude.

Gurzadyan explained that the success of the experiment depended on optical equipment of high precision and a unique three-axis platform on which the observatory was mounted which stabilized the system with an accuracy of several seconds of arc. Pointing was done initially by positioning the spacecraft within a few degrees and then two-axially orientating the instrument itself which had its own drive and automatic pointing system with an accuracy of 3–5 angular seconds.

Film obtained by Orion 2 was extracted and brought back to Earth by the cosmonauts, the equipment itself being left to burn up during re-entry with the jettisoned orbital module.

The Soyuz 13 cosmonauts soft-landed under conditions of low cloud and snow at 1150 hr. (Moscow time) on 26 December 1973 some 200 km. (124 miles) south-west of Karaganda, having completed 128 revolutions of the Earth.

Six months later came the attempt to restore the Soviet manned space-station programme. On 25 June 1974 a Proton-type D-1-H rocket put Salyut 3 into an initial orbit of 219–270 km. (136–168 miles) inclined at 51·65° to the Equator.

On 3 July Soyuz 14 with cosmonauts Colonel Pavel Popovich and Lieutenant-Colonel Yuri Artyukhin ascended from the cosmodrome at 1851 hr. GMT heading for a rendezvous with the station. After necessary orbital manœuvres the ferry reached the space-station at around midnight (Moscow time). At a range of 1 km. (0·6 mile) the cosmonauts switched to automatic control and took over manual control when the closing distance was down to 100 m. (328 ft) to achieve a soft-docking.

After checking on-board systems Artyukhin and Popovich entered the station at 0430 hr. (Moscow time) on 5 July.

Mission objectives on this occasion included: studies of the geological structure of the Earth's surface and atmospheric formations and phenomena 'with the aim of obtaining data for the solution of economic tasks'. There were also studies of the physical characteristics of space; bio-medical research on the effects of spaceflight on the human organism and work ability aboard the space-station; and testing of the station's improved design and on-board systems and equipment.

A week later it was explained that a replica of Salyut 3 was being used by Mission Control in which two scientists were carrying out the same procedures as their colleagues in orbit to confirm decisions and actions.

The two cosmonauts in space began by testing modifications which had been made to the station's basic design. These included a new set of three pivoting wing-like solar panels amidships which constantly faced the Sun to produce electricity regardless of the station's attitude. Each day the men used a boom-mounted TV camera to make external inspections of the space-station and its equipment and sent pictures to Earth.

Special gym equipment on board included a moving-belt device and a special sweatshirt with shock absorbers which placed a load on bones and muscles as the wearer ran on the spot.

The wardroom had food containers, a heating stove, a tape recorder, a radio, a small library, and a water tank. A varied cosmonauts' diet included spicy tomato sauce, wheat bread with ham, tinned minced bacon, fruit sticks, prunes and nuts, cottage cheese with black currants, honey cake, white coffee, vitamins, and a range of soups. Popovich had his favourite Ukrainian borsh and Artyukhin green soup.

Four portholes in the hull of the station provided a good view for making navigational measurements and still photography and filming were possible using hand-held cameras.

Soviet news sources made much of the economic potential of certain on-board experiments. Orbiting some 161 km. (100 miles) *lower* than the unoccupied Skylab space-station the cosmonauts were said to be photographing geological structures 'to determine areas likely to contain mineral deposits, land subject to salination, and the general condition of the soil'. They also studied the state of glaciers with the object of forecasting their movements. First observations reported were of central Asian republics and the Pamir Mountains.

In another experiment an RSS-2 spectrograph was used to study possible effects on the Earth's thermal environment and changes in climate caused by aerosol particles – particles of smoke and dust in the Earth's atmosphere.

A large solar telescope, it was disclosed, could be carried in a conical, unpressurized part of the station. The operator stood on a small platform at the base of the instrument and directed the telescope from a small control panel.

The fact that this instrument *faced the Earth* during long periods of the flight of Salyut led to speculation in the West that it was, in fact, a long-focus reconnaissance camera. The Russians insisted, however, that the compartment housed a solar telescope and that, to operate it, the station was rolled through 180°. For still photography of the Earth four aircraft-type cameras were mounted in the 'floor' and 'ceiling' of the station.

The Soyuz 14 cosmonauts were reported to have tested a system for the regeneration of water under conditions of protracted spaceflight; also a 'microbiological cultivator' for the further study of bacteria under weightlessness.

Throughout the flight careful watch was kept on the radiation environment by unmanned satellites and dosimeters incorporated in Salyut itself. One reason why the Russians chose to make their manned flights at low altitude was that the Earth's magnetic field deflects the worst of the radiation.

The mission by any standards was an outstanding success. After storing experimental results, film packs, and log books in the command module, the cosmonauts undocked from Salyut 3 at 0030 hr. (Moscow time) on 19 July 1974. Their command module soft-landed at 1521 hr. (Moscow time) about 140 km. (87 miles) south-east of Dzhezkazgan within 2,000 m. (6,560 ft) of the designated target.

Meanwhile, the station had been left to operate automatically under ground control with some of the systems closed down. Periodically its orbit was adjusted upward to prevent orbital decay in readiness for the next boarding party.

The expected launching came at 2258 hr. (Moscow time) on 26 August when cosmonauts Lieutenant-Colonel Gennady Sarafanov and flight engineer Lev Demin set off in Soyuz 15. Major-General Vladimir Shatalov, head of the cosmonaut group, described their mission as 'a continuation of the work of improving the transport ship and orbital station'.

By 1700 hr. (Moscow time) on 27 August, Soyuz 15 had completed 12 revolutions of the Earth following manœuvres which had made the orbit 254–275 km. (158–171 miles) inclined at 51·6° to the Equator.

Then, before midnight, Soviet Mission Control had their first indication of a fault in the control of the ferry. The surprise news that Soyuz 15 was returning came at 0800 hr. (Moscow time) on 28 August. The *Tass* announcement stated: 'Having approached

Salyut 3 many times, the spacecraft is being prepared to return.'

Western observers conjectured that the cosmonauts, having been forced to control the craft manually, had run low on fuel. Contributing to this theory was a report from Geoffrey Perry of the Kettering radio tracking group in England who had picked up strong signals from Salyut 3 as it passed over at about 0430 hr. BST on 28 August. He also made a visual sighting of the two space vehicles and found that Soyuz was some 17 sec. ahead of the station. The craft had been expected to dock around midnight.

The cosmonauts soft-landed in darkness at 2310 hr. (Moscow time) on 28 August some 48 km. (30 miles) south-west of Tselinograd, Kazakhstan. The *Novosti* press agency reported (without really explaining anything) that 'under the second day programme the cosmonauts made experiments to perfect the technique of piloting the spacecraft in different flight regimes. . . . Soyuz 15 approached Salyut 3 many times. The cosmonauts controlled the functioning of all the ship's on-board systems and made observations of the stages of approach, and they inspected the station when they approached it.' Of the night landing, it was said that this was part of a pre-planned search and rescue exercise.

On 5 September 1974 Salyut 3, having completed 1,162 revolutions of the Earth by noon (Moscow time), was continuing to fulfil the set flight programme in an automatic regime with all systems functioning normally. The orbit at this time ranged between 260 and 286 km. (161 and 178 miles).

Meanwhile, NASA officials were being pressed to get to the truth behind the docking failure because of any possible repercussions affecting the safety of the forthcoming Apollo-Soyuz Test Project. The Russians insisted that the problem had no connection with ASTP docking systems which were quite different.

Major-General Vladimir Shatalov during an ASTP training session at Houston, Texas, finally set the record straight on 9 September. He explained that the Soyuz 15 mission had been curtailed because of the failure of a fully automatic and remotely controlled rendezvous and docking system. During several attempts to achieve automatic docking with the station, the Soyuz automatic system caused overlong burns of the reaction control when the range was down to 30–50 m. (100–165 ft).

The experiment, he revealed, was related to a future capability of docking an unmanned supply craft with a manned Salyut space-station as a means of replenishing consumables. It had never been intended that the Soyuz 15 cosmonauts should occupy the station for experiments.

On 26 September *Tass* announced that the programme of work aboard Salyut 3 'has been carried out in its entirety'. It was then disclosed that on 23 September Salyut 3 had ejected a data capsule which had been recovered in the Soviet Union.

On 25 November the station, still unmanned, had completed 2,462 revolutions of the Earth. Several times Mission Control had used the on-board rocket engine to lift the orbit which then ranged between 247 and 293 km. (153 and 182 miles).

An official communiqué stated that tests of on-board systems were continuing. Opportunity was being taken to check the station's aerodynamic characteristics in its low orbit, and experiments were being carried out with on-board navigational systems. The characteristics of radio aids and thermal-regulation systems were also being studied.

The statement also revealed that the station had been permanently orientated on the Earth for five months and that observations were being made of 'areas of the Earth's surface by means of a television system for obtaining data on the Earth's natural resources in the interests of the national economy'.

At 1800 hr. (Moscow time) on 25 December 1974 Salyut 3, having operated for six months – double the intended duration – had completed 2,950 orbits. In this period more than 400 scientific and technical experiments were carried out, more than 8,000 commands were sent to the station for conducting over 200 operations, and 70 television and 2,500 telemetric sessions were held. The station's attitude control thrusters had been fired about 500,000 times and the electrical power system had generated some 5,000 kW/hr.

After necessary closing-down operations had been carried out, on 24 January 1975, Salyut 3 was made to re-enter the atmosphere over the Western Pacific and was consumed by friction.

Already, however, a replacement was orbiting the Earth. Launched from Tyuratam on 26 December 1974, Salyut 4 entered an orbit of 219–270 km. (136–168 miles) inclined at 51·6° to the Equator. The tracking ship *Akademik Sergei Korolyov*, which had remained on station near the east coast of Canada following the 6-day mission of Soyuz 16 (page 144), once again was used as the Atlantic communications link.

It was not long before Western tracking stations detected that this would be a different type of mission. By 1400 hr. (Moscow time) on 6 January 1975 the station had completed 180 revolutions of the Earth, having been manœuvred by ground control into a higher orbit of 343–355 km. (213–221 miles) with a period of 91·3 min.

This was a more logical orbit for long-duration space-station operations requiring less expenditure of fuel to overcome the effects of residual air drag. An orbit of this type had previously been flown by an unmanned spacecraft, Cosmos 613. This was thought to have been a test of an unmanned Soyuz ferry without solar 'wings'; launched on 30 November 1973 it remained in space for 60 days apparently with basic systems closed down in simulation of a docking mission involving a Salyut.

This made the earlier low-orbit Salyut stations requiring regular injections of thrust more reminiscent of high-resolution reconnaissance satellites (e.g. the USAF 'Big Bird'). The importance of the low-altitude application to the Soviet Union was underlined by the need to develop an unmanned Soyuz specifically for the task of replenishing consumables and returning film and data tapes to Earth.

Soyuz 17 with cosmonauts Lieutenant-Colonel Alexei Gubarev and Georgi Grechko was launched in pursuit of Salyut 4 at 0043 hr. (Moscow time) on 11 January 1975. After entering a preliminary orbit the craft was manœuvred in two stages to match the orbit of the space-station at approximately 350 km. (217 miles) altitude. According to *Novosti*, 'on-board automatic search and approach systems picked up the station when it was at a distance of about 4,000 m. (13,120 ft). The propulsion system was switched on and the closure rate increased to up to 12 m./sec. (39·4 ft/sec.).

From a distance of 100 m. (328 ft), operations to bring the craft closer to and dock with the station were handled by the crew.'

This was to be a wide-ranging scientific mission. Use was made of a Filin X-ray telescope to observe the star fields for bursts of radiation which might fix the position of pulsars and neutron stars. For this purpose Salyut 4 had an external platform with X-ray detectors linked in parallel with a small optical telescope. Any bursts of X-rays registered by the counters could thus be tied to a specific celestial body. Early telemetry tapes obtained from this experiment included X-ray emissions from the region of the Crab Nebula.

During the Salyut 4 mission a solar telescope housed in the non-pressurized 'funnel' in the large section of the station (page 235) was used to obtain spectra of the Sun in the UV region.

A whole series of genetic, embryological, physiological, and biological experiments were carried out with insects, micro-organisms, and higher plants.

Other research tasks were designed to study the physical nature of active solar processes and the Earth's ionosphere in regions of the electromagnetic spectrum which are beyond the reach of ground-based instruments.

The cosmonauts used biological apparatus of the 'Oasis' type in which to grow higher plants, observed the growth of chlorella in a nutrient solution and replanted micro-organisms in a special cultivator. Explaining the workings, Soviet scientist Alexander Kamin said it was a kind of 'cosmic factory' producing albumen and oxygen and served to lay the basis for a future space industry when scientists and engineers would remain in space for long periods. The experimental biological system operated automatically. All the cosmonauts had to do was put in a quantity of chlorella and ciné-cameras filmed the process.

Each working day the men of Salyut 4 ended with physical exercises to tone up the circulatory system and muscles weakened by the absence of gravity. The exercises included pedalling a bicycle ergometer fixed to the floor and 15 min. on a moving belt 'running track' (see *Frontiers of Space* (3rd edition) .

On 7 February 1975 the mission became the third longest

manned spaceflight, exceeding the 28 days set by the first Skylab boarding party. Later that day the radio tracking group at Kettering picked up voice reports from the space-station indicating that the cosmonauts were preparing to return.

Two days later at 0908 hr. (Moscow time) Soyuz 17 separated from Salyut 4 to make a successful re-entry, landing in difficult weather conditions at 1403 hr. (Moscow time) some 110 km (68 miles) north-east of the town of Tselinograd in Kazakhstan. Winds gusted to 71 km./hr. (44 m.p.h.), cloud height was 240 m. (800 ft) and visibility 5,000 m. (16,404 ft).

SOVIET TRACKING NETWORK

Code	Centre	Location
YeVP	Yevpatoria	Crimea
TBL	Tbilisi	West of the Caspian
DZhS	Dzhusaly	South-east of Tyuratam
KLP	Kolpashevo	North of Novosibirsk, on the Sputnik/Vostok ground track
ULD	Ulan Ude	Near Irkutsk
USK	Ussuriysk	Near Vladivostok
PPK	Petropavlovsk	Kamchatka

Ground stations are supplemented by tracking ships (typical locations: Gulf of Honduras and off Nova Scotia) linked by Molniya communications satellites. Mission Control for manned space operations at Kalinin.

The Soviets waited nearly two months before launching the second boarding party and when it came there was a major malfunction. Cosmonauts Colonel Vasily Lazarev and Oleg Makarov had just taken off in their Soyuz spacecraft on 5 April when the launch vehicle's top stage developed a fault sending the vehicle off course. Immediately the automatic abort system went into operation. Soyuz separated from the rocket stage, discarded its orbital and service modules, and pilot and main parachutes were deployed in sequence (see *Missiles and Rockets*, pp. 80–1). The command module made an emergency landing some 1,609 km. (1,000 miles) south-west of Gorno-Altaisk in Western Siberia.

As far as is known this was the first launch abort ever carried out involving a manned spacecraft. The same cosmonaut team that had flown the critical Soyuz 12 mission, testing the modified spacecraft after the Soyuz 11 tragedy – were recovered alive and well.

CHAPTER SEVEN

A MEETING IN SPACE

Imagine a spaceship marooned in orbit unable to return to Earth. Its crew are injured or in danger of suffocating as precious oxygen runs out. What are the chances of another nation sending up a rescue team to render medical aid or bring them home?

Until Russia and America got down to planning the Apollo-Soyuz Test Project (ASTP) the chances were slim indeed. Not only did the craft have different docking mechanisms but their space crews breathed different atmospheres at different pressures. To make them compatible involved years of dedicated engineering effort.

The idea of linking Apollo and Soyuz spacecraft in a joint exercise sprang from informal talks between Dr Thomas O. Paine, a former NASA Administrator, and the veteran Soviet space scientist Academician Anatoly Blagonravov in New York in April 1970.

To the great satisfaction of the astronautical community, it was taken up enthusiastically in the spirit of the détente when former President Nixon and Prime Minister Kosygin signed a space pact in Moscow on 24 May 1972.

If the project succeeded, and the political environment continued to improve, there was the opportunity to embark upon major collaborative projects. Part of the ASTP agreement ensured that all future manned spacecraft – including America's re-usable Space Shuttle – should be capable of docking together and transferring crews.

Vehicles which could dock with one another in Earth-orbit could lead to the construction of major space-stations built on the modular principle (see the companion volume *Frontiers of Space*). Not only could the international community exploit the space environment for economic benefit but a major manned space-station could become the gateway to the Moon and the planets.

The ASTP experiment, in itself a relatively small step in terms of the space commitment, had larger implications. After the Cold War era of the fifties and sixties, it was a visible sign of a major shift of policy within the Kremlin. Every few months large space teams – astronauts/cosmonauts, engineers, and officials – commuted between Houston in Texas and Zvezdnoy Gorodok (Star Town) on the outskirts of Moscow for training in each others' spacecraft and simulators. To ensure absolute safety during the mission it was agreed that the Russians speak English and the Americans speak Russian. Lettering on the consoles of Soyuz and Apollo appeared in both languages.

NASA selected for the Apollo astronaut team USAF Brigadier-General Thomas P. Stafford, forty-four; Donald K. 'Deke' Slayton; and Vance D. Brand. Stafford had previously flown in Geminis 6 and 9 and the Moon-orbiting Apollo 10. Slayton would be fifty-one years old when the mission took place. One of the original seven Mercury astronauts selected in 1959, he was grounded because of a slight heart irregularity which had since been corrected. Brand, forty-four, an astronaut since 1966 also had never flown in space.

The back-up crew comprised USN Captain Alan L. Bean; USN Captain Ronald E. Evans; and US Marine Corps Lieutenant-Colonel Jack R. Lousma.

Russia's Soyuz team included no fewer than four crews: prime crew commander was the veteran cosmonaut Alexei A. Leonov, a Lieutenant-Colonel in the Soviet Air Force – the first man ever to space-walk from an orbiting spacecraft (page 133). His companion Valery N. Kubasov had carried out the first welding experiment in space during the flight of Soyuz 6 in 1969.

Reserve crews were: (1) Air Force Colonel Anatoly V. Filipchenko, the former command pilot of Soyuz 7, and Nikolay N. Rukavishnikov, who flew in Soyuz 10 as test-engineer; (2) Air Force Major Vladimir Dzhanibekov and civilian Boris Andreyev; and (3) Air Force Captain Yuri Romanenko, and civilian Aleksander Ivanchenko.

To make the link up possible several fundamental modifications had to be made to both Apollo and Soyuz. Not only did the craft have different docking systems but their crews breathed different atmospheres at different pressures. Soyuz had an oxygen-nitrogen atmosphere at full sea-level pressure; Apollo had pure oxygen at about one-third that value.

Any attempt by the spacemen to change from one craft to another without first acclimatizing to the different atmospheres would create problems similar to those faced by deep-sea divers in ascending too rapidly from the ocean floor. They could suffer 'the bends' owing to the formation of nitrogen bubbles in their blood.

Making the ships compatible involved the design, by American and Russian engineers, of a special airlock/docking module to be launched with the Apollo spacecraft. Built by the Space Division of Rockwell International in Downey, California, this tube-like structure about 3·05 m. (10 ft) long by 1·52 m. (5 ft) diameter was the passageway airlock that separated the living quarters of the two craft after they had docked.

The Apollo itself was a modified version of the command/service module used on manned lunar exploration missions, one that had actually been built for the lunar programme but not used. It now carried additional propellants for the reaction control (orientation and stabilization control) system, heaters for temperature control, and equipment required to operate the docking module.

Modifications to Soyuz included an improved life-support system, additional air regulation blocks, a new androgynous docking system and a modernized motor control system. Impulse beacons and signal lights were added to aid Apollo in locating and docking with Soyuz.

Collaboration had gone so well that by early 1974 the participants had agreed a tentative flight plan:

On 15 July 1975 at about 1520 hr. (Moscow time) Russia's Soyuz spacecraft crewed by Leonov and Kubasov blasts off from the Tyuratam cosmodrome to achieve an orbit at 230 km. (140 miles) altitude inclined at 51·8° to the Equator.

Approximately 7½ hr. later the Apollo spacecraft crewed by Stafford, Slayton, and Brand ascends from Cape Canaveral in Florida. The airlock/docking module (pages 90–4) is carried by the same 720,000 kg. (1·6 million lb) thrust Saturn IB rocket in a compartment immediately behind the spacecraft. Once in orbit the Apollo CSM separates from the rocket's final stage, turns around and docks with the module pulling it free. A similar technique was used to extract the Apollo lunar module on the way to the Moon (page 190).

The initial orbit of 150–170 km. (93–105 miles) is circularized at 230 km. (140 miles) as Apollo closes with Soyuz.

Docking with Soyuz occurs some 50 hr. after the launch of Apollo, the extended docking mechanism of the latter being mated with the retracted docking mechanism of the Russian ship illustrated on pages 90–1. At initial contact, the three guides mounted on the guide ring (3) of each mechanism align the spacecraft in the proper attitude for capture. Further movement of the spacecraft toward each other operates three capture latches (6) to complete the initial docking position of the two spacecraft. The Apollo docking mechanism is then retracted, drawing the structural rings (5) together. Eight structural latches automatically lock with the eight structural latches of the other mechanism. When the structural latches are locked, the two spacecraft are in the final docked position.

At this stage rubber seals of both mechanisms (4) are compressed into each other. The internal pressure is then increased in the area of the seals and pressure checks are made to ensure there are no leaks.

With the living quarters of the American and Russian spacecraft now set nose to nose, with the passageway airlock module between them, transfer from one to the other is relatively simple. First, the Soyuz cabin pressure is reduced to about 530 mm. (10 p.s.i.a.) and oxygen content raised. Then, in the case of an American transfer, two crewmen enter the airlock module and close the door behind them. Gradually the atmosphere changes to match that of the other ship, whereupon the adjoining hatch door is opened and the men pass through.

Lightheartedly, the Russians even announced the menu of the celebration dinner that would await their Western colleagues. Their guests would be offered a choice of hot soups from cuisines of different peoples of the USSR: Ukrainian beetroot-and-cabbage soup, a piquant Georgian mutton broth, and Russian sorrel and spinach soup. The second course included veal, chicken, pâté, ham, and sausage meat. Soups, served in tubes, were heated in special ovens in the cabin. Tins and food cubes were firmly attached to the table with special cords so that they did not float away. For dessert there were prunes with nuts, cake, and juices.

Soyuz and Apollo were to remain docked for approximately two days during which period each member of the crew would take turns to visit the other ship. The combined US-USSR crews would make joint experiments and radio and TV reports would be sent from both ships.

After final undocking from the Soyuz – following redocking exercises – at approximately 99 hr. 15 min. Soyuz ground elapsed time, each spacecraft was to conduct independent activities. The Soyuz would continue in orbit for some 43 hr. after separation, landing at about 142 hr. 30 min. Soyuz GET in Kazakhstan. The Apollo would continue in orbit for about 3½ days landing in the Pacific near Hawaii.

Throughout the entire flight it was arranged that Mission Control centres of both countries would remain in radio communications contact with each other as well as with their spacecraft.

The project cost the United States approximately $250 million. What it cost the Soviet Union was not revealed but the seriousness with which the Russians viewed the exercise can be judged from the number of ASTP spacecraft and launch vehicles committed to the programme. Two Soyuz spacecraft with crews were prepared for launch day.

In preliminary tests two spacecraft were launched unmanned to check modifications to the basic Soyuz:

Cosmos	Launch Date	Orbit (km)	Inclination to Equator°	Notes
638	3 April 1974	258–274 *km. (160–170 miles)	51·78	Capsule recovered after 9·9 days.
672	12 Aug. 1974	227–238 *km. (141–148 miles)	51·76	Capsule recovered after 5·9 days

* After orbital manœuvres.

Then came a full dress rehearsal. Soyuz 16, which ascended from Tyuratam at 1240 hr. (Moscow time) on 2 December 1974, put into orbit the first reserve crew – cosmonauts Filipchenko and Rukavishnikov. Identical with the actual ASTP craft, the aim was to test the ability to reduce pressure in the Russian ship, bringing it closer to the low-pressure oxygen atmosphere of Apollo. In effect this meant reducing pressure inside the Soyuz from 760 mm. (14·7 p.s.i.a.) to approximately 530 mm. (10 p.s.i.a.) and raising oxygen content from 20 to 40 per cent.

A 20 kg. (44 lb) movable ring device, mounted on the nose of Soyuz 16, replaced Apollo in the docking exercise making it possible to place the same loads on the Russian craft in simulation of docking. After the experiment the test apparatus was discarded in orbit.

The manœuvres executed by the spacecraft followed closely the flight plan of the actual mission and the cosmonauts returned, as planned, on 8 December, landing some 300 km. (186 miles) north of Dzhezkazgan.

The actual mission also was a complete success. Launched at 1220 hr. GMT on 15 July, the Russian Soyuz 19 arrived in the pre-planned rendezvous orbit at 225 km. (140 miles) altitude inclined at 51·76° to the Equator. Apollo approached Soyuz and docked with it at 1615 hr. GMT on 17 July. After Stafford and Slayton had acclimatized in the airlock module the hatch door was opened and the celebrated 'space handshake' between Stafford and Leonov, the two commanders, was televised to the World.

During the two days the ships were together, crew exchanges and joint experiments were carried out. Finally, on 19 July, the men returned to their own craft and hatches were re-sealed. After a separation and re-docking exercise (with Soyuz the active partner) the ships separated for the last time to make individual experiments (pages 90–2).

It was the last spaceflight that Americans were scheduled to make in a throwaway rocket. When astronauts returned to space it would be in their re-usable Space Shuttle.

CHAPTER EIGHT

SPACECRAFT IN FOCUS

Vostok (pages 17, 19, 20–6)

Certain technical features of the world's first manned spacecraft were kept secret for four years after the pioneer single orbit performed by Major (later Colonel) Yuri Gagarin on 12 April 1961. We now know that the spacecraft had two main sections, a spherical re-entry capsule of 7·5 ft diameter and an instrument compartment which included a retro-rocket system.

The capsule itself was comprehensively clad in an ablative material for protection of the cosmonaut against frictional heating and swathed in strips of metal foil as a means of reflecting solar heat in orbit. A rocket ejection seat for the cosmonaut slid into the capsule on rails, with access through a large circular hatch. The seat rails were inclined to the horizontal ensuring ejection of the cosmonaut at an upward angle. Exit from the vehicle was possible by blowing explosive studs securing the escape hatch immediately behind the cosmonaut's head.

Of the six cosmonauts who made orbital flights in Vostok spacecraft, Gagarin alone remained inside the capsule for the landing. Owing to the capsule's high sinking rate beneath its recovery parachute, the normal procedure was to eject at low altitude following re-entry for separate parachute recovery.

Whereas the American Mercury spacecraft was stabilized by gas-jets during re-entry, Vostok's spherical capsule had no such refinement. After the vehicle had been ejected from orbit by rocket braking, four tensioning bands holding the capsule to the instrument section were released. The latter, having served its function, was allowed to plunge into the atmosphere and burn up. Meanwhile, the spherical re-entry body containing the cosmonaut slowly assumed its correct attitude as it encountered the Earth's atmosphere.

Orientation was achieved by the simple procedure of weighting the capsule forward of the geometric centre so that, as air pressure built up, it slowly swung round into the correct re-entry attitude to take the brunt of frictional heating on the thickest part of the heat shield. In this attitude the cosmonaut, lying on his ejection seat, was subjected to re-entry deceleration 'chest-to-back' in the approved manner. It seems probable that early Soviet ballistic warheads were developed on this principle, creating confidence in the technique for manned experiments.

The world received a first glimpse of 'the Vostok spacecraft' at the Tushino Air Display in July 1961, when a full-size representation was displayed beneath an Mi-6 helicopter. In fact, this was a replica of the rocket's final stage complete with nose-fairing which hid all detail of the spacecraft within; and to add further confusion the rocket module had been fitted with a tail annulus supported by eight fins.

Although at the time Western observers thought the tail annulus had been added to stabilize the exhibit beneath the helicopter, the fact that the feature subsequently appeared on official Soviet commemorative stamps and in a documentary film lends support to the view that it was, in fact, a mischievous distortion. At all events, when a Vostok was at last shown without its nose-fairing at the Economic Exhibition in Moscow in April 1965, there was no aerodynamic tail. It was only then that one saw that the re-entry capsule was, in fact, a sphere.

I have questioned leading Soviet space officials on more than one occasion without receiving a satisfactory answer, being advised merely to: 'Believe the Moscow exhibit!' Since the re-entry capsule configuration was kept secret for four years after Gagarin's pioneer flight, one can only conclude that the Russians wanted to keep the West, and particularly America, ignorant of essential design features.

Providing the capsule with an all-enveloping heat shield meant that special provision had to be made for three portholes and the three hatches required for recovery parachutes, ejection seat, and equipment inspection. While the hatches were covered in the same ablative material as the remainder of the capsule, the portholes, made of refractory glass, were deeply recessed in the outer covering.

Within the capsule the cosmonaut was fully protected against emergencies which might otherwise prove fatal. If anything went wrong at lift-off he could escape by blowing the hatch and ejecting in the seat. As the seat rails were inclined in the capsule he would gain height before the parachute opened. Similarly he was able to eject from the capsule after returning from orbit.

The ejection seat, fitted with two ejection rockets, had a detachable back, straps to secure the cosmonaut, and a parachute. In case re-entry occurred in an unscheduled area, possibly over the sea, the seat also contained emergency rations of food and water, radio equipment, and an inflatable dinghy.

Within the cabin the arrangement of instruments and equipment allowed the cosmonaut to perform his flight duties with minimum effort. Instruments were grouped according to their functional role. The console to the left of the cosmonaut had instruments for regulation of temperature and air humidity. There were also radio equipment controls and controls for orientation of the spacecraft during orbital flight.

Vostok's attitude control was assisted by two solar sensors, one automatic and one manual, which allowed the craft to be orientated in a given direction. This, of course, was vital when it came to firing the retro-rockets to reduce orbital speed and initiate re-entry. Vostok was turned about its centre of mass by steering jets fed from compressed gas storage bottles. Switches and buttons on the side panel of the console enabled the cosmonaut to fire the retro-rocket at any point of the orbit.

Immediately facing the cosmonaut was an instrument panel with a clock and counters indicating the number of orbits completed. However, the main instrument was an Earth-globe which revolved synchronously with the spacecraft's movement round the world. This made it possible for the cosmonaut to determine his geographical position at any time and also to predict the area into which his craft would descend should it be necessary to initiate re-entry at any particular point of the orbit. Mounted beneath this panel was a television camera, and below that a porthole incorporating the 'Vzor' optical orientation device. With this instrument the cosmonaut could line up the spacecraft with the Earth's horizon and therefore obtain manual control of the orientation system. The optical system installed in the viewing port comprised two annular mirror-reflectors, a light filter, and a lattice glass. Light rays travelling from the line of the horizon struck the first reflector and, passing through the glass of the porthole, reached the second reflector which directed them through the latticed glass to the cosmonaut's eyes. If the spacecraft's attitude in respect to the vertical axis was correct, he would see the horizon in the form of a circle in his field of vision. Then, using a control handle, the cosmonaut could orientate the craft to bring the line of the horizon into view as a concentric circle. This would be proof of his correct positioning, ensuring that the direction of the Earth's horizon coincided with the course plotted on the glass.

The control handle for manual orientation of the spacecraft was situated on the right-hand side of the cabin together with radio equipment and a food container.

Normally the cosmonaut reclined on his ejection seat with the visor of his helmet open and the suit ventilated with cabin air. In the event of the cabin becoming depressurized, the helmet visor would be closed, pressurization of the suit following automatically. Under such emergency conditions, a reserve supply of compressed oxygen and air would allow time for the cosmonaut to make contact with a ground station, take a decision, select a landing site, and perform an emergency re-entry. The manual control facilities in the cabin allowed the cosmonaut himself to pitch the Vostok at a backward inclined angle prior to firing the retro-rocket. In Vostok it was possible to fire the retro-rockets both automatically (when automatic landing systems were employed) and manually using the controls previously mentioned.

The cabin which had a nitrogen-oxygen atmosphere was designed to maintain normal atmospheric pressure and oxygen content, a carbon dioxide content of not more than 1 per cent, a temperature of 15 to 22° C, and a relative humidity of 30 to 70 per cent. Regeneration of the air, including absorption of CO_2 and water vapour, was assured by means of 'highly active chemical compounds'.

The procedure was automatically controlled. Should the amount of oxygen drop and CO_2 content increase, a sensor gave a signal which adjusted the supply. Similarly, if there was an excess of oxygen, the amount fed into the cabin was automatically adjusted. Humidity was controlled in the same way.

In the event of contamination of the cabin air by harmful admixtures resulting from the function of the human body and the work of the instruments, filters purified the atmosphere. There was also a special temperature control system employing a constant-temperature coolant to conduct heat away from the cabin. The coolant flowed through the temperature control system to a liquid-gas radiator. On the outside of the spacecraft was an automatic radiative heat exchanger, with a system of shutters, which served to maintain the temperature of the cooling agent at the required level.

Food, water, air, and electrical supplies in the Vostok were sufficient for a flight lasting up to 10 days. Had the retro-rocket failed to operate or any other fault occurred preventing the spacecraft's return at the appointed time, the orbit was deliberately chosen to ensure that re-entry would occur within this period as the natural result of atmospheric retardation.

Communications equipment included a Signal radio transmitter operating on a frequency of 19·995 Mc/s for purposes of ground tracking. Two-way radio contact with Earth stations was maintained using short-wave radio on 9·019 and 20·006 Mc/s and ultra-short-wave on 143·625 Mc/s.

The FM channel allowed reliable contact with ground stations up to distances of 930 to 1,240 miles. On short-wave it was possible to maintain communications with ground stations in the Soviet Union over the greater part of the orbit. A tape-recorder was provided in the radio-telephone circuit for recording the cosmonaut's comments for later transmission when Vostok passed over appropriate ground stations.

Vostok's nose-fairing was an interesting feature. It backed on to the attachment ring of the instrument module and final stage and was rocket-ejected after the launch vehicle had penetrated the lower atmosphere. At first sight it appears a dangerous procedure to encase the cosmonaut's re-entry capsule in a massive shroud. What would happen if it failed to separate or something went wrong with the rocket at lift-off?

In fact, the Soviet designers overcame the problem very neatly by sculpting a large cut-out in the side of the fairing through which the cosmonaut could eject after blowing the escape hatch. However, clearly ejection could not be left too late or, despite the protection afforded by his pressure suit, the cosmonaut would perish upon re-entering the Earth's atmosphere.

Many tests of escape systems were made from geophysical rockets and unmanned spacecraft using dogs as test-subjects; there were also high-altitude drops of human subjects from aircraft and balloons.

Much of the work concerned with testing parachutes and ejection seats for the Vostok was performed by the late Colonel Pyotr I. Dolgov. Dolgov was rumoured to have lost his life in a Soviet space mishap. In fact, he died on 3 November 1962, while making a high-altitude parachute jump from the stratospheric balloon Volga. The balloon was at a height of some 25,000 m. (82,021 ft). Major E. Andreev, who jumped first, landed safely; but Dolgov died on the way down.

Voskhod (pages 19 and 27)

Few details of the Voskhod 1 spacecraft were immediately released by the Soviet authorities. There was accommodation for three men with two seats abreast and one slightly forward. Pressure suits were not worn but kept for emergency use in a locker. Unlike Vostok, the spacecraft did not embody ejection seats.

Although not actually demonstrated, Voskhod was stated to have manœuvre capability and an environmental control system capable of supporting the crew for 1 month. There were two retro-rocket systems for ejecting the capsule from orbit, main and reserve. A retro-rocket landing system served to cushion the touchdown as the parachute-supported capsule came within a few feet of the ground, reducing the vertical velocity before impact to perhaps 3 to 4 ft/sec.

Once again the cabin atmosphere was nitrogen-oxygen at sea-level pressure.

An article in *Pravda* on 29 March 1965 described the internal layout of the Voskhod 2 cabin. Two comfortable armchairs, upholstered in white, stand next to each other; close to hand are instrument panels with numerous tumbler switches and luminous dials. Directly overhead are two instrument panels, one of them the control panel for the airlock chamber, equally convenient for either cosmonaut to operate. On the right-hand side, above Belyaev's seat, is the spacecraft's control panel. A long black handle operates the manual orientation system. Just beyond this panel, beneath a transparent safety catch, are the instruments used for re-entry. A red metal hood covers a small black button marked 'Descent TDU' – the Russian initials for retro-engine group. This, of course, is the button that fires the retro-rocket to start re-entry into the atmosphere after the spacecraft has been pitched backwards at the correct angle by the orientation system.

On the other side of the cabin, at upper left, is an instrument panel containing an Earth-globe. As the spacecraft moves round the Earth so the globe revolves indicating the cosmonauts' position at any moment. Adjacent on the same panel are an electric clock, and several needle indicators. Alongside are a signalling board and the optical orientation system control panel.

Ciné and television cameras look down from above. At the extreme left is the airlock hatch. Frosted bulbs are switched on in the airlock chamber and a ciné-camera is installed in a corner. The airlock control panel installed in the cabin is duplicated in the airlock chamber so that, in case of need, the cosmonaut can assume control of the airlock by simply pressing a button. Space-equivalent conditions are then created in the airlock chamber and, at a signal from the cabin, the second hatch swings open to the void of space.

From the photographs of Leonov emerging from Voskhod 2, the airlock was just over 6 ft long by 3 ft diameter. In order to capture movements of the cosmonaut photographically, a number of external directional antennae provided continuous TV transmissions during 2 to 3 r.p.m. roll manœuvres. The improved

TV system – supplying pictures both to the spacecraft cabin and ground stations – employed 625 scan lines and 25 frames per sec. VHF signals were transmitted from the spacecraft on 143·625 Mc/s, ultra-short-wave on 17·365 and 18·035 Mc/s.

In submitting particulars of the Voskhod 1 mission for ratification by the International Aeronautical Federation (FAI), the Soviet authorities gave the aggregate thrust of the multi-stage launch vehicle as 650,000 kg. (1,433,250 lb) using a total of seven engines. This compares with 600,000 kg. (1,323,000 lb) of the Vostok launcher employing six engines.

Soyuz (page 139)
The three-man Soyuz spacecraft, introduced in 1967, consisted of three basic modules: *Orbital module*, located on the forward end, used by the crew for work and rest during orbital flight; a docking unit was located on the central axis. Overall length: 2·65 m. (8·7 ft); diameter 3·35 m. (7·3 ft). Weight, 1,122 kg. (2,700 lb). *Descent module*, with main controls and crew couches, used by crew during launch, descent, and landing. Overall length: 2·2 m. (7·2 ft). Weight, 2,802 kg. (6,200 lb). *Instrument (or service) module* at rear, with sub-systems required for power, communications, propulsion, and other functions. Electrical power supplied by chemical storage batteries and large extensible, wing-like, solar cell panels of 14 $m.^2$ (150 ft^2) area. Life-support system maintained oxygen-nitrogen cabin atmosphere at about 760 mm. (14·7 p.s.i.a.). Combined habitable volume (all compartments) about 9 $m.^3$ (317·8 ft^3).

Attitude control of the Soyuz spacecraft depends on small hydrogen peroxide thrusters located in the rear of the service module. The system includes 14 × 10 kg. (22 lb) and 8 × 1 kg. (2·2 lb) thrust units. (A back-up system includes 4 × 10 kg. (22 lb) and 4 × 1 kg. (2·2 lb) thrusters.)

Six H_2O_2 10 kg. (22 lb) thrusters in the command module are employed for re-entry.

Main propulsion of Soyuz depends on a primary engine of 417 kg. (917·4 lb) thrust installed on the central axis of the instrument module. A 411 kg. (904·2 lb) thrust back-up engine with a single chamber has twin exhaust nozzles, emerging on either side of the main chamber nozzle. Propellants are a type of nitric acid and a hydrazine derivative. Exhaust ports for the H_2O_2 pump turbines are also located at the back of the engine bay.

For Soyuz 10 and subsequent missions an internal transfer facility was provided in the orbital module to permit crew transfer between docked craft, following unmanned testing within the Cosmos programme.

After the Soyuz 11 depressurization tragedy in 1971, the third cosmonaut position in all Soyuz spacecraft was replaced by additional life-support equipment, allowing a two-man crew to wear lightweight spacesuits for additional (back-up) protection during take-off, descent, and landing. Previously, crew-members on non-EVA assignments had worn light woollen clothing with flying helmets and headsets.

ASTP Soyuz (pages 90–2 and 252)

A modified version of the standard spacecraft was used for the joint US/Soviet Apollo-Soyuz Test Project (ASTP). This included changes to the life-support system to permit reduction of cabin pressure from 760 to 530 mm. of mercury (14·7 to 10 p.s.i.a.) to provide a 'transitional atmosphere' for purposes of crew transfer including an increase in oxygen content from 20 to 40 per cent.

The ASTP craft also carried a new docking unit and improved power supply and attitude control systems. Impulse beacons and signal lights were fitted to help Apollo in the rendezvous manœuvre.

Soyuz space-station ferry (page 236)

Another Soyuz variant, introduced in 1973, dispensed with solar 'wings' and relied for power supply entirely upon chemical storage batteries. This spacecraft was designed specifically to fly two cosmonauts to a space-station, dock with it, and return the command module. When docked with the station it was possible to re-charge chemical batteries from the solar 'wings' of the station itself.

Cosmos 573 – which made a two-day flight unmanned between

15 and 17 June 1973 – appears to have tested the stripped-down configuration. Early manned tests were made by Soyuz 12.

A spacecraft of this type – Soyuz 15 – tested an automatic system for docking with Salyut which in the event malfunctioned (page 242). The experiment, according to Soviet sources, was part of a project to develop an unmanned supply craft for the purpose of replenishing consumables. The requirement was particularly important in the case of low-orbit stations which had consequently to expend rocket fuel to sustain the orbit against the effects of air drag. This implied a space-station having duplicate docking facilities which neither Salyut 3 nor Salyut 4 possessed. However, the automatic docking system was also applicable to the remote assembly of space-station modules. As early as August 1974 Major-General Shatalov had described the new Soyuz as a 'universal spacecraft both for independent manned flight and for the assembly of complex objects in orbit . . . also for carrying freight and crews to scientific stations'. Ships of this type, he said, 'undoubtedly will become assembly sites for large space-stations to be set up in orbit'. The Russians, it seemed, were on their way to developing a tele-controlled space tug.

Salyut space-stations (page 236)

The Salyut 1 space-station consisted of three basic cylindrical elements with the docking unit located forward on the central axis. Adjacent to the docking unit was a short cylinder of about 2 m. (6·56 ft) diameter, which expanded into two additional cylinders of approximately 2·9 m. (8·8 ft) and 4·15 m. (13·6 ft) respectively. All these sections were hermetically sealed and pressurized. Aft of the pressure bulkhead of the largest cylinder were propellant tanks (in the form of a sphere and cone) and beyond that a cylindrical propulsion bay of about 2 m. (6·56 ft) diameter. At front and rear were sets of extensible wing-like solar panels similar to those used on the Soyuz ferry.

Overall length with Soyuz docked was about 23 m. (70 ft) and the combined habitable volume nearly 100 m.^{3} (3,531 ft^3), the orbital module of the ferry being used as sleeping quarters.

Going aboard Salyut, one first entered an interconnecting tunnel which contained part of the 'astrophysical apparatus' and several control panels. This led to a hatch beyond which was the main working compartment with seats for two cosmonauts mounted on a low platform facing the hatch. The cosmonauts faced control consoles and at their sides were command and communications equipment of the type used in Soyuz spacecraft.

Three other work positions allowed the crew to carry out various research tasks, and next to one of these was a porthole; also in this section was life-support equipment including air regulation units and filters. Further aft were more vital systems and bio-medical apparatus. An array of scientific equipment included the Orion 1 astrophysical observatory and the Oasis 1 closed-cycle biological apparatus.

Salyut 3

Salyut 3 (page 236) embodied a number of changes to the basic design. They included a new set of three pivoting wing-like solar panels amidships which were permanently orientated on the Sun to generate electricity regardless of the station's attitude in space. The two sets of fixed wing-like solar arrays fore and aft on Salyut 1 were deleted.

Salyut 4

Further details of the internal design of the basic station were forthcoming with the launch of Salyut 4. The 'working module' comprised two cylinders of 2·9 m. (9·42 ft) and 4·15 m. (13·62 ft) diameter with a conical intersection. This served not only as a 'workshop' but also as a 'study room, dining room, bedroom, drawing room and gym'.

In the smaller section were the station's central control panel, a table with built-in food heaters, and another work station for controlling life-support systems and some scientific experiments.

The larger section had sleeping accommodation along the port and starboard sides, food storage refrigerators, a movable track and an ergometer 'bicycle' for physical training (designed and assembled by the Likhachov motor works in Moscow). There were two airlocks for ejecting encapsulated refuse that burnt up in the atmosphere, and another work station for controlling experiments.

On either side of the working module were removable decorative panels concealing frames which supported various service equipment. The 'walls' of the module were painted in light colours (green, yellow, and blue) and the 'floor' dark grey.

McDonnell Mercury (pages 30–5, 38)

America's first manned spacecraft was a bell-shaped capsule designed to provide an astronaut with complete experience of spaceflight, including rocket launching, weightlessness (first in sub-orbital and subsequently in orbital flight), re-entry, and landing. The configuration was determined primarily by conditions which the capsule would experience under frictional heating when re-entering the Earth's atmosphere.

The astronaut's cabin included a form-fitting couch and an environmental control system providing 100 per cent oxygen at a pressure of 5·5 p.s.i. under space conditions for both cabin and pressure suit. There were two individual control circuits for cabin and suit respectively which normally operated simultaneously for about 28 hr. The astronaut could isolate the suit circuit from the cabin circuit by closing the face-plate of his helmet. The astronaut's pressure suit was not normally inflated except in the case of a failure in the cabin circuit.

The pressure shell of the cabin was double skinned with 0·010 in. titanium, with forward and rear pressure bulkheads also of 0·010 in. titanium. Separated from the inner pressure cone by top-hat stringers was an outer heat-protective skin, made largely of beaded shingles of 0·016 in. René 41 nickel alloy. Between the pressure shell and the outer skin was ceramic-fibre insulation.

The capsule's cylindrical neck section (housing main and reserve landing parachutes) was formed from 0·22 in. beryllium shingles; the adjacent antenna-cone was of Vycor topped by 0·031 in. René 41.

The ablative heat shield of orbital capsules was a glass-fibre reinforced laminated plastic; beryllium was used in the case of sub-orbital capsules where re-entry conditions were less severe. Shields were made detachable so that, following re-entry and release of the main landing parachute, they could be dropped down approximately 4 ft to deploy a perforated impact skirt of rubberized glass-fibre which became a pneumatic cushion on impact with the ocean. After 'softening' the touchdown, it rapidly filled with water to become a sea-anchor maintaining the capsule in an upright floating position.

Two-way communications between the astronaut and ground stations were maintained during flight. Equipment included a voice radio, a receiver for commands from ground stations, and a radio tracking beacon. The astronaut's bio-medical reactions were recorded in flight. Instruments continuously monitored the internal and external environment. Engineering, medical, and scientific data were telemetered to ground receivers.

Four control systems ensured safety of the craft under conditions of flight and re-entry. Eighteen thrusters located in the cylindrical and conical sections of the capsule released bursts of superheated steam for control in pitch, yaw, and roll. Six of these were in a self-contained manual-mechanical system.

Thruster operation depended on passing hydrogen peroxide over a catalyst which decomposed it into steam and oxygen. The peroxide was fed through shut-off valves to the control solenoids under helium pressure and thence to the thrust nozzles.

Mercury's 'autopilot' was the Automatic Stabilization and Control System (ASCA). Its primary function was to detect any unstable motions of the capsule and bring about corrective torques through controlled bursts of thrust; the system could also be programmed to orientate the capsule with the heat shield forward in the path of flight.

A Rate Stabilization and Control System (RSCS) included gyros which sensed rate changes to bring about capsule orientation.

The astronaut was able to assume control by means of the Manual Proportional Control System (MPCS). This had the separate set of six thrusters. The three-axis hand controller worked proportional throttle valves above the thrusters via mechanical linkages; wrist action produced response in yaw, fore and aft movement in pitch, and lateral movement in roll.

Lastly, the Fly-by-Wire System (FBWS) enabled control

column movements to work the throttle valves of the main thrusters electrically. Thus, on 'automatic' and Fly-by-Wire, twelve nozzles were used, those acting in pitch and yaw being mounted in the capsule's cylindrical neck; they delivered a fixed burst of thrust at 1 lb or 24 lb. Roll-control nozzles, located near the base of the capsule at the heat-shield end, gave either 1 lb or 6 lb thrust. The low-thrust nozzles served to bring about minor attitude corrections while the high-thrust nozzles were applied for orientating the capsule and pitching it up in the re-entry attitude. The six manually operated nozzles had variable thrust, the pairs acting in pitch and yaw exerting 4 to 24 lb and the two roll nozzles 1 to 6 lb.

The long curved tubular tank for the hydrogen peroxide thrusters was fitted between the rear pressure bulkhead of the cabin and the heat shield. Stowage for the folded pneumatic landing skirt was in the adjacent compartment.

Attached to the capsule at lift-off was a tower-mounted escape rocket with three canted nozzles. If the main stage engines failed at lift-off or the vehicle deviated from course beyond prescribed limits, errors would be sensed by instruments which would signal the abort system. Immediately, the 50,000-lb thrust escape rocket would separate the capsule from the errant booster for parachute recovery. There was no separate ejection seat for the astronaut. In normal circumstances three 350-lb thrust solid rockets jettisoned the tower when the vehicle had cleared the lower atmosphere.

The retro-rocket pack (secured to the heat shield by metal straps) contained three 1,160-lb thrust solid rockets. With the capsule pitched up at 34°, these were ripple-fired to reduce speed by about 350 m.p.h. Following re-entry a 6-ft diameter conical ribbon-type drogue opened at 21,000 ft. At 10,000 ft this was followed by deployment of the 63-ft diameter ringsail landing parachute stowed in the capsule's cylindrical neck.

McDonnell Gemini (pages 36, 37, 41, 42, 45–59)

The Gemini spacecraft differed in several respects from its predecessor. Although shaped like Mercury it had approximately

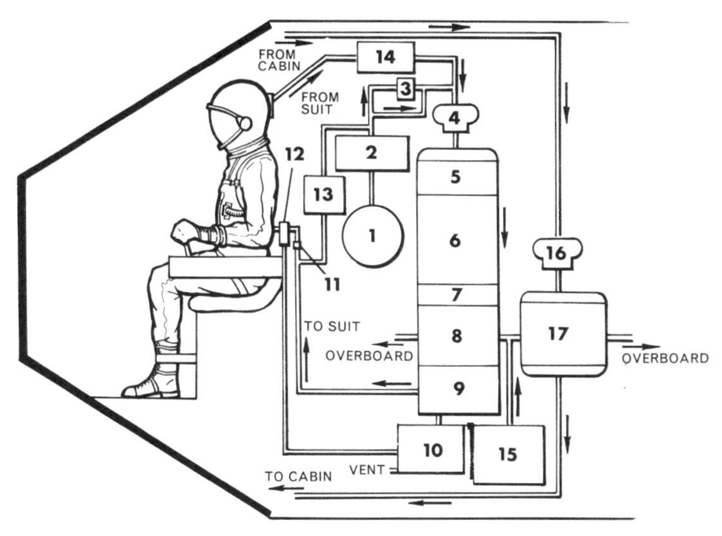

Mercury spacecraft: Environmental Control System (ECS). *Key:* 1. Oxygen; 2. Pressure reducer; 3. Regulator; 4. Fan; 5. Odour absorber; 6. Carbon dioxide absorber; 7. Filter; 8. Heat exchanger; 9. Water separator; 10. Condensate tank; 11. CO_2 sensor; 12. Condensate trap; 13. Emergency flow; 14. Debris trap; 15. Cooling trap; 16. Fan; 17. Heat exchanger.

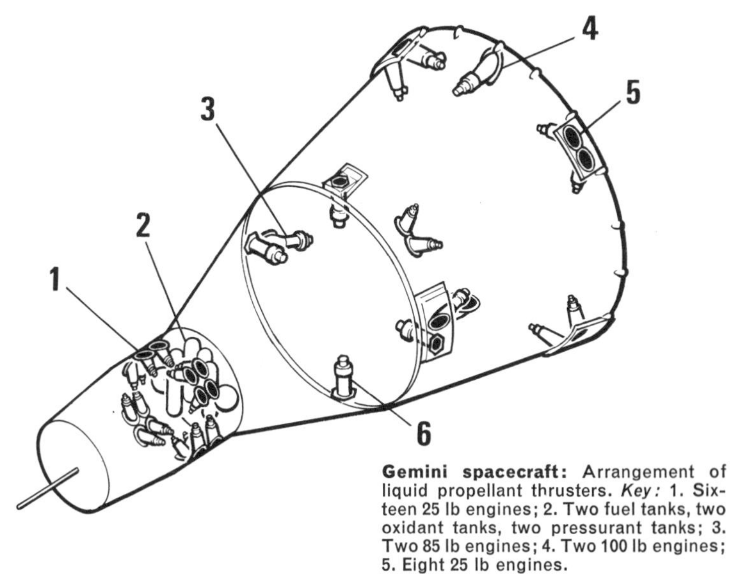

Gemini spacecraft: Arrangement of liquid propellant thrusters. *Key:* 1. Sixteen 25 lb engines; 2. Two fuel tanks, two oxidant tanks, two pressurant tanks; 3. Two 85 lb engines; 4. Two 100 lb engines; 5. Eight 25 lb engines.

Meals consumed aboard Gemini spacecraft were a great improvement over the paste-forms taken on earlier missions. There were two types of food, bite-sized and hydratable. This is the menu for astronauts James Lovell and Edwin Aldrin in Gemini 12; each was allowed 13 meals.

Day 1: Meal A	*Calories*
Applesauce	139
Sugar-frosted Flakes	139
Bacon Squares	180
Cinnamon Toast	56
Cocoa	190
Orange Drink	83
	787

Day 1: Meal B	
Pea Soup	220
Tuna Salad	214
Cinnamon Toast	56
Date Fruitcake	262
Pineapple-grapefruit Drink	83
	835

Day 1: Meal C	
Beef Pot Roast	119
Potato Salad	143
Cinnamon Toast	56
Chocolate Pudding	307
Brownies	241
Tea	32
	898

Day 2: Meal A	
Applesauce	139
Sugar-frosted Flakes	139
Bacon Squares	180
Cinnamon Toast	56
Cocoa	190
Orange Drink	83
	787

Day 2: Meal B	
Beef/Vegetables	98
Meat/Spaghetti	70
Cheese Sandwiches	158
Apricot Pudding	300
Gingerbread	183
Grapefruit Drink	83
	892

Day 2: Meal C	
Pea Soup	220
Tuna Salad	214
Cinnamon Toast	56
Date Fruitcake	262
Pineapple-grapefruit Drink	83
	835

Day 3: Meal A	*Calories*
Peaches	98
Strawberry Cereal Cubes	171
Sausage Patties	223
Cinnamon Toast	56
Orange Drink	83
Grapefruit Drink	83
	714

Day 3: Meal B	
Potato Soup	220
Chicken Salad	237
Beef Sandwiches	138
Butterscotch Pudding	311
Tea	32
	938

Day 3: Meal C	
Shrimp Cocktail	119
Beef and Gravy	160
Creamed Corn	105
Toasted Bread Cubes	161
Pineapple Fruitcake	253
Orange-grapefruit Drink	83
	881

Day 4: Meal A	
Fruit Cocktail	87
Toasted Oat Cereal	91
Bacon Squares	180
Ham/Applesauce	127
Cinnamon Toast	56
Orange Drink	83
Pineapple-grapefruit Drink	83
	707

Day 4: Meal B	
Shrimp Cocktail	119
Chicken/Gravy	92
Toasted Bread Cubes	161
Pineapple Fruitcake	253
Coconut Cubes	206
Orange-grapefruit Drink	83
	914

Day 4: Meal C	
Beef/Vegetables	98
Meat/Spaghetti	70
Cheese Sandwiches	158
Apricot Pudding	300
Gingerbread	183
Grapefruit Drink	83
	892

50 per cent greater internal volume. It comprised two principal sections, the re-entry module containing the pressurized crew compartment, and the adapter module.

The crew compartment was a pressurized vessel surrounded by an unpressurized volume in which various systems and fuel were housed. Major components were arranged in modular packages which could be replaced through easily removed external panels without opening the crew compartment or affecting systems already checked out, thereby reducing maintenance and check-out periods prior to lift-off.

Cabin pressurization with 100 per cent oxygen was limited from 5 to 5·3 p.s.i. above ambient by the cabin dual pressure regulating valve.

The basic structure reflected the constant effort made in spacecraft design to achieve optimum strength-to-weight ratios. Ring-frame stabilized stringers carried nearly all axial loads, principal metals being titanium and magnesium. The pressure shell itself had a fusion-welded titanium frame attached to side panels and fore and aft bulkheads formed of double thickness, thin-sheet titanium (0·010 in.) with the outer sheet beaded for maximum stiffness.

Following the practice in Mercury the outer surface of the re-entry module was covered by overlapping shingles for aerodynamic and heat protection behind the blunt heat shield. The beaded shingles of 0·016 in. René 41 were thermally isolated from the stringers. Between the pressure compartment and the outer skin was a layer of insulation.

The small end of the re-entry module, skinned with unbeaded beryllium shingles, was capped by a nose-fairing of reinforced plastic and glass-fibre laminate.

The spacecraft's ablative heat shield was based on a load-carrying sandwich of glass fibre. Two five-ply face-plates of resin-impregnated glass cloth were separated by a 0·65-in. thick glass fibre honeycomb core; an additional glass-fibre honeycomb was bonded to the convex side of the sandwich and filled with Dow-Corning DC-325 organic compound. This ablative substance was a paste-like material which hardened in standard atmosphere after being poured into the honeycomb form. Completing the shield on the periphery was a Fiberite ring.

Unlike Mercury, Gemini had no escape tower. Emergency escape – either during the ascent phase to more than 60,000 ft or after re-entry – was provided by ejection seats similar to those in modern military aircraft. There were two large hatches, one for each astronaut, which could be opened and locked open for escape, even under maximum dynamic conditions.

Each hatch had an observation window comprising three glass panes with an air space between each one. The command astronaut's window had two outer glass panes of 96 per cent silica glass and an inner pane of temper-toughened alumino-silicate glass. The co-pilot's window was constructed for optical clarity; instead of the alumino-silicate pane it had a 96 per cent silica glass increased in thickness from 0·22 to 0·38 in. Optical transmission capability of the three panes exceeded 99 per cent for observation and photography.

The astronauts' spacesuits were automatically pressurized with 100 per cent oxygen at 3·7 p.s.i. when the cabin was depressurized for extra-vehicular activity. The 23-lb suit worn by the command pilot had five layers: (1) a white cotton constant wear under-garment with pockets round the waist for bio-medical instrumen-

---

(Opposite)

The most ambitious extra-vehicular activity (EVA) experiment planned for a Gemini astronaut involved use of two modules, a chest pack Emergency Life Support System (ELSS) and a backpack Modular Manœuvring Unit (MMU). Together with the Gemini pressure suit these comprised the Astronaut Manœuvring Unit (AMU) shown on page 57, a system that is essentially a miniature human spacecraft. The experiment, which was not completed during the Gemini 9 mission, required the EVA astronaut to make his way from the Gemini cabin to the back of the adapter module with the use of hand-holds to don the MMU backpack. Pictures show the intended sequence of operations: 1. MMU stowage at back of adapter module; 2. Ejection of cover plate; 3. EVA astronaut extends hand-rail; 4. Astronaut transfers to adapter module; 5. Seats into MMU and fixes 100 ft tether extension; 6. Leaves with MMU backpack for manœuvre experiments adjacent to spacecraft. Results of the experiment which had to be curtailed are given in Chapter Two, page 179.

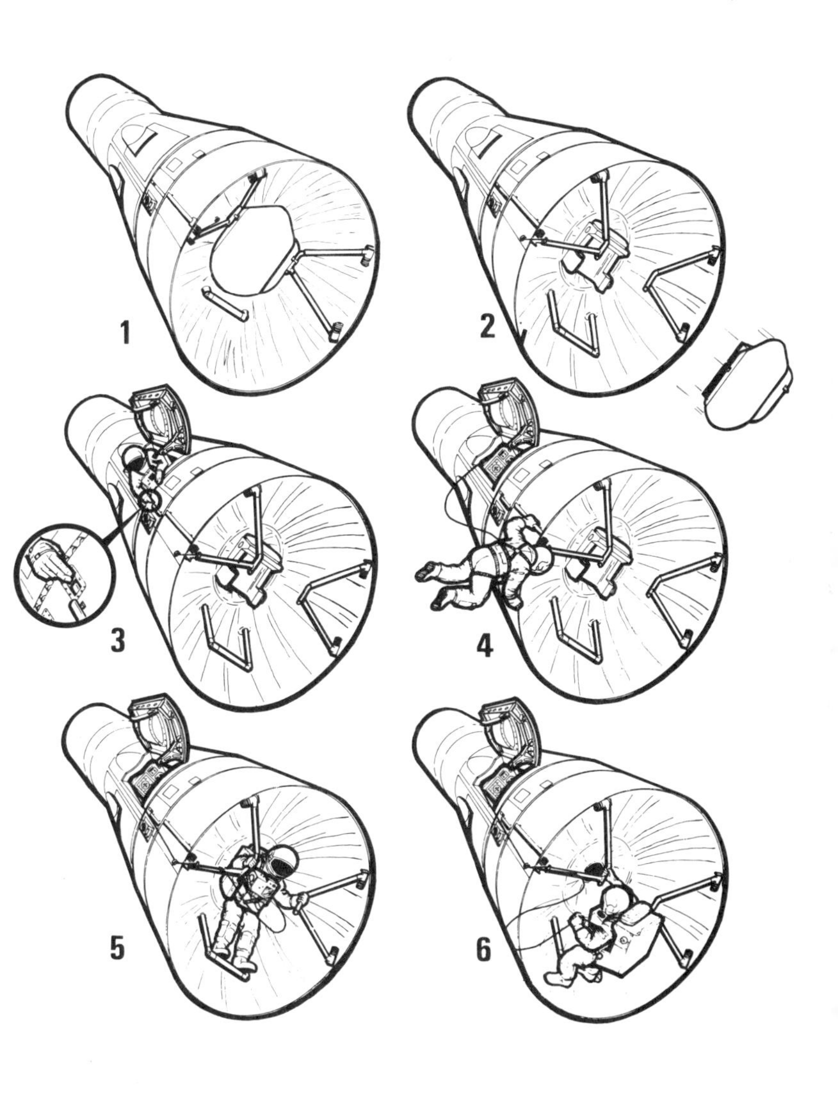
1
2
3
4
5
6

tation; (2) a blue nylon comfort layer; (3) a black neoprene-coated nylon pressure garment; (4) a restraint layer of nylon link net to restrain the pressure garment and maintain its shape, and (5) a white nylon outer covering.

The EVA suit, weighing 33 lb, had seven layers: (1) the white cotton undergarment with waist instrumentation pockets; (2) the blue nylon comfort layer; (3) the black neoprene-coated nylon pressure garment; (4) a restraint layer of Dacron and Teflon link net; (5) a thermal protective layer of seven layers of aluminized Mylar with spacers between each layer; (6) a micrometeoroid protective layer, and (7) a white nylon outer covering.

Toilet facilities aboard Gemini, as in all early manned spacecraft, were not particularly elaborate. Elimination of body waste involved attaching to the body a plastic bag with an adhesive lip. The bag contained germicide to prevent the formation of bacteria and gas. After use the bag was sealed with the adhesive lip and stowed in an empty food container, the contents being brought back for analysis. Urine was delivered into a fitted receptacle connected by hose to either a collection device or an overboard dump.

The spacecraft's adapter module consisted of separate retrograde and equipment sections which were jettisoned prior to re-entry. In the latter was the main power supply comprising two General Electric hydrogen-oxygen fuel-cells; five silver-zinc batteries were located in the equipment bay of the crew compartment for re-entry, landing, and post-landing operations.

Attitude control was obtained from small thrust units (25, 85, and 100 lb) using self-igniting monomethyl hydrazine and nitrogen tetroxide (in contrast to Mercury's hydrogen peroxide). In the retrograde section were six Orbit Attitude and Manœuvring System (OAMS) thrusters. Four of these allowed orbital translation, up, down, left, and right. Two had nozzles facing the re-entry module for rearward translation.

Ten OAMS thrusters in the equipment section gave control in pitch, yaw, and roll and provided forward translation in orbit. The re-entry module itself had two rings of eight thrusters for attitude control on the return to Earth.

Infra-red scanners, which sensed spacecraft roll and pitch attitudes with reference to the Earth's horizon, provided attitude information. When switched to automatic control, this system operated almost like an aircraft automatic pilot, leaving the astronauts free to direct their attention elsewhere. Alternatively, manual control in pitch, yaw, and roll could be obtained with the three-axis hand controller.

The rendezvous radar, installed in the small end of the spacecraft, had the function of measuring range, range rate, and bearing angle to the target vehicle. This enabled the crew to determine manœuvres necessary for rendezvous and docking. A transponder on the target vehicle received radar impulses and returned them to the spacecraft at a specific frequency and pulse width. Weight of the spacecraft radar was under 70 lb and the power requirement less than 80 W.

Gemini's inertial guidance system comprised a digital computer, the inertial platform electronics package, and a power supply. The computer was a general-purpose, binary, digital type capable of performing high-speed integrations, input-output conversions, arithmetic computations, and rendezvous timing functions. It had a non-destructive memory and automatically indicated any fault. A manual data insertion unit allowed the astronauts to place up to 100 messages into the computer, and served as a back-up for the digital command system. Three push buttons on a control panel allowed the astronauts to enter, cancel, or read out the displayed data.

The inertial measuring unit, operating in conjunction with the computer, measured all accelerations applied to the spacecraft. It depended on a stabilized platform containing three miniature integrating gyros and three pendulous accelerometers. The system served both as an attitude reference and a navigation reference during rendezvous, re-entry, and landing.

Heat-protective paints and coatings were responsible for the dark appearance of the re-entry module. The white finish on the adapter module, on the other hand, allowed maximum release of heat brought to the metal surface by coolant channels beneath the skin. This part of the spacecraft was made of circular aluminium-alloy frames, extruded magnesium-alloy stringers, and magnesium skin. The T-shaped stringers had a hollow bulbous section allowing the passage of liquid coolant, which transferred heat generated in the vehicle to the skin for radiation into space.

A cloth cover over the open end of the adapter module protected the internal equipment from solar radiation after separation from the launch vehicle. This was made of Vitron-impregnated glass cloth with vapour-deposited gold.

Recovery of the spacecraft from orbit was effected by firing four solid-propellant retro-rockets, each producing 2,500 lb thrust by 5·5 sec. and operating in series with overlap. Following re-entry, at about 60,000 ft, an 18-ft diameter ringsail parachute was deployed by a mortar; then, at 10,000 ft, an 84-ft diameter ringsail parachute inflated to provide stable descent towards the sea at a vertical velocity of 30 ft/sec.

The Lunar Project

The evolution of the spacecraft can be traced to many independent sources in many countries. Its theoretical basis was laid at the turn of the last century by Konstantin E. Tsiolkovsky, the father of Soviet cosmonautics, and Professor Hermann Oberth in Germany contributed hugely to the theory and practice of space travel in his classic work, *Die Rakete zu den Planetenraumen*, first published in 1923. Many other workers were active in this period; but the practical development of large liquid-propellant rockets, under Major-General Walter Dornberger and Dr Wernher von Braun at Peenemünde between 1938 and 1945, turned the dream into potential reality.

The BIS Spaceship (page 74)

Credit for the first engineering analysis is due to the pre-war Technical Committee of the British Interplanetary Society. In 1938 the Society set out to examine the practical requirements for a vehicle capable of carrying three men to the Moon and back. The results of a design-study were published in 1939.* Although

* See *Journal of the British Interplanetary Society*, January and July, 1939.

MANNED SPACEFLIGHT CHRONOLOGY: THE FIRST FIVE YEARS

Astronaut	*Spacecraft (Name or Call-sign)*	*Country*	*Launch Date*	*Flight Duration*
Gagarin	Vostok 1 (Swallow)	USSR	12.4.61	1 hr. 48 min.
Shepard	Mercury MR-3 (Freedom 7)	USA	5.5.61	15 min. 22 sec.
Grissom	Mercury MR-4 (Liberty Bell 7)	USA	21.7.61	15 min. 37 sec.
Titov	Vostok 2 (Eagle)	USSR	6.8.61	25 hr. 18 min.
Glenn	Mercury MA-6 (Friendship 7)	USA	20.2.62	4 hr. 55 min.
Carpenter	Mercury MA-7 (Aurora 7)	USA	24.5.62	4 hr. 56 min.
Nikolayev	Vostok 3 (Falcon)	USSR	11.8.62	94 hr. 22 min.
Popovich	Vostok 4 (Golden Eagle)	USSR	12.8.62	70 hr. 57 min.
Schirra	Mercury MA-8 (Sigma 7)	USA	3.10.62	9 hr. 13 min.
Cooper	Mercury MA-9 (Faith 7)	USA	15.5.63	34 hr. 20 min.
Bykovsky	Vostok 5 (Hawk)	USSR	14.6.63	119 hr. 6 min.
Tereshkova	Vostok 6 (Sea Gull)	USSR	16.6.63	70 hr. 50 min.
Feoktistov Komarov Yegorov	Voskhod 1 (Ruby)	USSR	12.10.64	24 hr. 17 min.
Belyaev Leonov	Voskhod 2 (Diamond)	USSR	18.3.65	26 hr. 2 min.
Grissom Young	Gemini GT-3	USA	23.3.65	4 hr. 53 min.
McDivitt White	Gemini GT-4	USA	3.6.65	97 hr. 56 min.

Astronaut	*Spacecraft (Name or Call-sign)*	*Coun-try*	*Launch Date*	*Flight Duration*
Cooper Conrad	Gemini GT-5	USA	21.8.65	190 hr. 56 min.
Borman Lovell	Gemini GT-7	USA	4.12.65	330 hr. 35 min.
Schirra Stafford	Gemini GT-6	USA	15.12.65	25 hr. 51 min.
Armstrong Scott	Gemini GT-8	USA	16.3.66	10 hr. 42 min.
Stafford Cernan	Gemini GT-9	USA	3.6.66	72 hr. 21 min.
Young Collins	Gemini GT-10	USA	18.7.66	70 hr. 46 min.
Conrad Gordon	Gemini GT-11	USA	12.9.66	71 hr. 17 min.
Lovell Aldrin	Gemini GT-12	USA	11.11.66	94 hr. 34 min.

Spacecraft times certified to the FAI by the Soviet authorities for homologation of flights as world records were: Vostok 3, 94 hr. 9 min. 59 sec.; Vostok 4, 70 hr. 43 min. 48 sec.; Vostok 5, 118 hr. 56 min. 41 sec.; Vostok 6, 70 hr. 40 min. 48 sec.

For subsequent manned missions see pages 144 (Soyuz) and 228, 288–9 (Apollo).

there were obvious shortcomings, particularly in terms of propulsion and post-lunar re-entry into the Earth's atmosphere, the study was remarkably perceptive of modern spacecraft techniques.

The cylindrical launch vehicle, 100 ft tall by 20 ft diameter, weighed 1,000 tonnes. Its propulsion system comprised 2,490 solid-propellant motors arranged as a honeycomb in steps or banks. The pressurized cabin was situated in the nose and clusters of rockets were built up layer by layer beneath it, so that the spent motors in the first stage dropped off and the next layer fired, and so on, until the vehicle reached escape velocity. Ignition was governed by an automatic trajectory programming device and liquid-propellant vernier motors were to give fine control of velocity and manœuvre. It was proposed to conduct launching from a flooded caisson sunk in a high-level lake near the Equator.

More layers of rockets were assumed to be used in braking against the Moon's gravitational pull and for the return flight. A unique feature of the design was that the vehicle employed extendable landing legs permitting a vertical touchdown after rocket braking. Therefore the spacecraft was placed in the correct attitude for re-launching from the leg supports. Small rocket motors were incorporated for mid-course correction of the ship's path, and for manœuvring prior to touchdown on the Moon.

From these few details it can be seen that as early as 1939 big advances had been made from the theoretical and visual conceptions of the 1920s. The BIS study envisaged the use of inertial navigation instruments. Such devices have since reached a high state of perfection for submarines, missiles, and aircraft. Another novelty – a 'coelostat' – was made and demonstrated. This, the first instrument intended specifically for a spacecraft, was an optical device designed to provide a stationary view of the heavens from a vehicle which was being rotated to provide the crew with a sensation of weight induced by centrifugal force.

Attention was paid to the physiological needs and safety of the crew, including the use of contour couches. A double-walled pressure cabin was proposed for good thermal insulation and to minimize risk of puncture by meteorites. Although the problem of re-entry into the Earth's atmosphere following the return from the Moon received little attention, a jettisoned heat-resisting nose-cone was shown protecting the vehicle from frictional heating during the ascent from Earth. Many of these ideas have since been applied in missiles and spacecraft.

The drawing on page 74 is a reconstruction in colour of an original drawing by the late R. A. Smith dated 1947. It represents a re-evaluation of the pre-war BIS Lunar Spaceship in terms of liquid-propellant technology. The vehicle was designed to lift off from the Moon using the leg-supported base section as a launch platform. A quarter of a century later this was to become the technique for re-launching an actual manned spacecraft from the lunar surface in project Apollo.

Apollo Command Module (pages 63, 64, 66, 73, 76–89).
This module housed the three-man Apollo crew and was the only part of the vehicle that returned from the lunar mission. It had the shape of a cone approximately 12 ft high and 12·8 ft across the base. The habitable volume was 218 ft^3. Fully loaded at lift-off it weighed approximately 11,000 lb.

The double-walled structure consisted of an inner pressure shell and an outer heat shield separated by structural stringers. A micro-quartz fibre between the walls provided thermal insulation. The outer structure was a three-piece heat shield (covering top, sides, and blunt end of the capsule), constructed of brazed honeycomb stainless steel to which was bonded an epoxy resin ablative coating. Thickness of the ablative material varied according to the expected aerodynamic heat distribution over the vehicle.

The inner pressure shell, made of aluminium honeycomb bonded between sheets of aluminium, was the primary load-carrying structure. An access tunnel, capped by a pressure cover, extended from the crew compartment to the apex. The space round the tunnel just below the apex was divided into four sections by stiffeners; these housed recovery parachutes, pyrotechnics, and electronic equipment.

The cabin was pressurized with 100 per cent oxygen at 5 p.s.i. ambient by the Environmental Control System (ECS). Other systems included S-band communications, guidance and navigation, and attitude control.

The Reaction Control System (RCS) provided control in pitch, yaw, and roll after separation from the service module for re-entry into the Earth's atmosphere. Twelve units mounted near the base of the module delivered 93 lb of thrust through ablative chambers. There were two sets of thrusters for each control axis. Fuel was monomethyl hydrazine and oxidant nitrogen tetroxide. Thrust could be applied automatically from the ASCS or independently by astronaut manual control. At launch an escape tower was fitted for separation of the command module under the action of the Emergency Detection System (EDS). The escape rocket would be operated during powered flight if there was catastrophic loss of thrust in the main-stage engines, or if excessive angular rates of pitch, yaw, or roll developed. These actions, automatically sensed by the EDS, would either fire the escape system directly or provide data for the crew to exercise manual control of abort procedures.

The wing-like canard surfaces (see drawing) would be opened by explosives 11 sec. after the escape rocket fires. They would serve to stabilize the command module blunt end forward prior to deployment of the drogue parachutes. There was also a pitch motor to deflect the craft into a recovery trajectory.

The 'Q-ball' at the top of the escape tower contained pressure sensors for determining flight angles of attack and dynamic pressure after lift-off.

It was also possible to use the engine of the service module as an escape system, independently of the tower-mounted rocket. In this case the command and service modules would separate following the abort manœuvre.

The landing system catered for pad abort, high altitude abort (up to 70,000 ft), and normal re-entry after flight. Following descent to 25,000 ft above sea-level, a barometric pressure switch fired an explosive charge to jettison the apex heat shield over the parachute compartment. Two seconds later, two 13-ft diameter nylon conical ribbon drogue parachutes were mortar-deployed for stabilization and deceleration of the command module. Drogues were reefed for 8 sec., then fully opened. At approximately 10,000 ft three pilot parachutes were deployed, also by mortar. Each pulled one of the main 83·5-ft ringsail parachutes from its container. Main parachutes opened in reefed condition, providing deceleration without excessive opening shock. After 8 sec. reefing lines were cut automatically and main parachutes fully inflated for final landing at approximately 27·5 ft/sec. Each main parachute weighed 127 lb including canopy, riser, and deployment bag. Total weight of the landing system was approximately 540 lb.

In a Third Report on the work of the Board investigating the fire in Apollo spacecraft AS 204 (pages 185–7), certain recommendations were made mainly affecting modifications to the Apollo command module.

The recommendations were:

> That combustible materials now used be replaced wherever possible with non-flammable materials, that non-metallic materials that *are* used be arranged to maintain fire breaks, that systems for oxygen or liquid combustibles be made fire-resistant, and that full flammability tests be conducted with a mockup of the new configuration.
>
> That a more rapidly and more easily operated hatch be designed and installed.
>
> That on-the-pad emergency procedures be revised to recognise the possibility of cabin fire.

In addition the Board drew attention to

> a number of components, sub-systems, techniques and practices which it felt could be improved to increase crew safety and mission reliability. These included findings in the environmental control system, solder joints, location of wiring, electrical equipment qualification and design, and the development of checkout procedures.

The Board did not recommend a change in the use of pure oxygen in the spacesuit for either pre-launch or orbital operations, nor did it recommend that cabin atmosphere for operations in space should be changed from 5 p.s.i. pure oxygen. However, it did recommend 'that the trade-offs between one and two-gas atmospheres be re-evaluated' and that pressurized oxygen no longer be used in pre-launch operations.

Apollo Service Module (pages 63, 64, 76–87)

This 12·9-ft long cylindrical vehicle supported the command module; the fully fuelled weight was approximately 55,000 lb. The structure was formed of 1-in. thick panels of aluminium honeycomb sandwich construction supported by six radial beams. Tension ties attaching the service module to the command module were released explosively to separate the modules prior to re-entry.

At the base of the module was the spacecraft's main propulsion unit used for mid-course correction, braking the vehicle into lunar orbit, and for the return flight. The engine, which was gimballed by two electrically operated servo actuators, delivered a maximum thrust of 21,900 lb. Propellant was self-igniting unsymmetrical dimethyl hydrazine and oxidant nitrogen tetroxide fed by helium pressure. There were four main propellant tanks.

The same propellant combination (in separate tanks) was used to power four independent and identical sets of four 100-lb thrusters of the RCS. All sixteen thrusters were mounted outside the service module, being fed from internal tanks under helium pressure.

The sets of four thrusters providing control in pitch, yaw, and roll were mounted with 90° spacing round the service module, and thrusters were mounted in a 90° relationship to each other. Thrust could be applied: (1) to force main engine propellant into the lower end of the tank for engine restarting; (2) for making minor orbital and mid-course manœuvres, and (3) for docking with the lunar module.

After Apollo 13

Engineers worked night and day to ensure that the Apollo 13 mishap – which so nearly took three lives – would not be repeated. Modifications to the command and service modules involved reducing to a minimum the use of Teflon, aluminium, and other materials potentially combustible in the presence of high-pressure oxygen, and keeping those that remained well clear of possible ignition sources. All electrical wires were sheathed in stainless steel and the quantity probe (indicating the amount of LO_2 remaining in a tank) was changed from aluminium to stainless steel. The fuel-cell oxygen supply valve was redesigned to isolate Teflon-coated wires from the oxygen.

Warning systems on board the spacecraft and at Houston Mission Control were also modified to provide more immediate and visible warnings of system anomalies. In addition, a third oxygen tank was added to the service module to avoid operations in low oxygen conditions, which made possible the removal of unsealed fan motors in the tanks.

Plans to increase stay-time on the Moon reached fruition with the introduction of J-series spacecraft in Apollo 15. The craft launched into orbit – CSM and LM – were nearly 2,903 kg. (6,400 lb) heavier than vehicles involved in the first Moon landing.

This involved modifications to the Saturn 5 launch vehicle and a revision to the departure trajectory to inject the spacecraft into Earth-orbit some 18·5 km. (11·5 miles) below the nominal altitude of previous missions. The five F-1 engines of the S-1 first stage were up-rated by increasing the flow rate of propellants and four of the eight retro-rockets were removed from the stage; all four ullage rockets were deleted from the S-11 second stage.

Modifications to spacecraft included:

(a) An increase in SM oxygen and hydrogen to 502 kg. (1,107 lb) – 176 kg. (389 lb) more than in Apollo 13 for electrical and environmental control systems.

(b) Addition to the SM of a Scientific Instrument Module (SIM) bay in a sector which was previously vacant.

(c) Other major modifications were reserved for the lunar module. In the descent stage a major rearrangement of the four outer quads increased supplies of oxygen and water and raised the electrical life of the vehicle. In Quad 1 space was provided for stowage of the Lunar Roving Vehicle (LRV) in a folded configuration. In Quad 4 (which had previously housed the Modularised Equipment Stowage Assembly (MESA)) were a new 50 kg. (111 lb) water tank, a waste container, and an extra gaseous oxygen tank and a gaseous oxygen module. A new MESA – containing sample containers, tool pallet, batteries for personal life-support systems, and a cosmic ray detector – was carried external to the quad. The four descent engine propellant tanks were each increased in length by 8·6 cm. (3·4 in.) providing an extra 521 kg. (1,150 lb) of fuel and oxidant. In compensation for the extra weight, the descent engine was allowed to burn for a longer period (changes were also made to the combustion chamber liner to reduce erosion and the expansion skirt was modified). Only relatively minor changes were made to the ascent stage. These included the addition of a plumbing system to the waste container in the descent stage, and a new method of stowing the redesigned pressure suits. The new suit had greater flexibility for leg and arm movements; and there were increased supplies of oxygen, water, and electricity to achieve longer EVA periods.

Overall the various modifications to the CSM and LM raised the combined weight of Apollo 15 spacecraft to 46,862 kg. (103,311 lb), an increase of nearly 2,359 kg. (5,200 lb) over Apollo 14.

Apollo Adapter Module (page 63)
This 28-ft long, 3,900-lb, tapered section served as the interstage fairing between the third stage of the Saturn 5 launch vehicle and the Apollo spacecraft. It also garaged the lunar module. The eight panels hinged back like petals to allow the lunar module to be pulled free after Apollo had been injected into the translunar orbit. The panels were of aluminium honeycomb sandwich construction 1·75 in. thick; four had linear explosive charges installed at the panel junctions.

Apollo Lunar Module (pages 65, 69–72, 78–82)
The spacecraft, designed to land two astronauts on the Moon after release from the Apollo parent craft in lunar orbit, had two main components; a *descent stage* incorporating the landing gear and an *ascent stage* containing the pressurized crew cabin. Earth launch weight was 33,205 lb in the case of Apollo 11.

Constructed largely of aluminium alloy, the structural shell of the ascent stage had an external composite layer of insulation and a thin aluminium skin affording the astronauts thermal and micrometeoroid protection. The outer skin and the inner pressure shell were approximately 3 in. apart. The cabin, a 92-in. diameter cylinder, was stiffened by 2-in. deep circumferential frames spaced approximately 10 in. apart and located between the structural skin and the thermal shield. The compartment had two triangular cabin windows in the front-face bulkhead, an overhead docking window on the left side, a forward hatch, controls and displays, and astronaut support equipment.

Directly behind the cabin was a smaller compartment containing the docking hatch, ECS, and equipment stowage. A removable cover provided access to the 3,500-lb thrust ascent engine plumbing and valves.

The upper docking tunnel at the top of the ascent stage was used for astronaut transfer to the command module of the Apollo parent. The hatch below the centre control consoles was used for leaving and entering the spacecraft on the lunar surface and for extra-vehicular transfer of crew and equipment in space.

Hatches had pre-loaded elastomeric silicone compound seals mounted in the lunar module structure. When the hatch was closed, a lip near the outer circumference of the hatch entered the seal, ensuring a pressure-tight contact. Both hatches opened inwards and normal cabin pressurization was used to force the hatch into the seal. To open either hatch, it was necessary to depressurize the cabin by means of a dump valve.

The ECS maintained pressure, temperature, and relative humidity for a maximum of 48 hr.* The cabin could be refilled four times with 100 per cent oxygen, and the astronauts' portable life-support system, five times. With a cabin pressure of 5 p.s.i. maximum leakage rate was 0·2 lb/hr.

Aft of the midsection pressure bulkhead were two gaseous oxygen tanks for the ECS (which supplied the cabin atmosphere), two helium tanks for ascent-stage main-propellant pressurization, and inverters and batteries for electrical power supply. Propellant tanks were located on both sides in the non-pressurized area. They held fuel and oxidant for the ascent engine and fuel, oxidant and helium for the RCS. Two ECS water tanks were located overhead in the ascent stage, with two gaseous oxygen storage tanks in the aft equipment bay.

The descent stage of the spacecraft, also built in aluminium alloy, had an inner structure with an outer composite layer of insulation and a thin aluminium-alloy skin. The outer skin formed a modified octagon shape round the descent stage. Two pairs of transverse beams arranged in a cruciform, together with an

* Later extended.

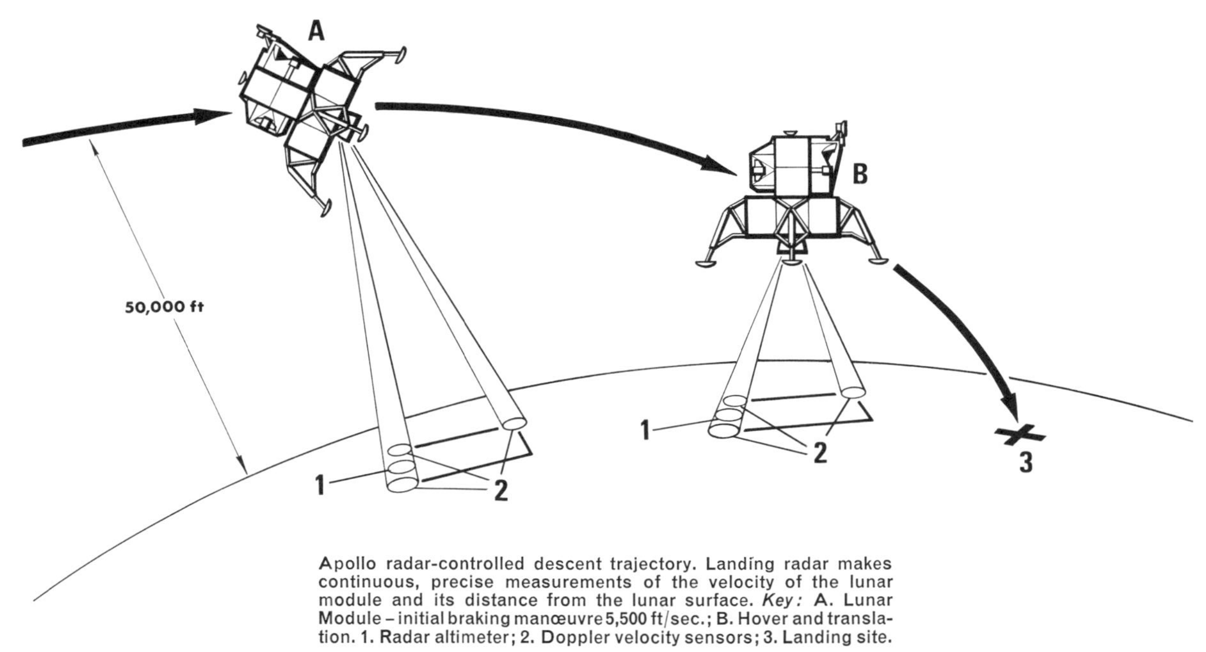

Apollo radar-controlled descent trajectory. Landing radar makes continuous, precise measurements of the velocity of the lunar module and its distance from the lunar surface. *Key:* A. Lunar Module – initial braking manœuvre 5,500 ft/sec.; B. Hover and translation. 1. Radar altimeter; 2. Doppler velocity sensors; 3. Landing site.

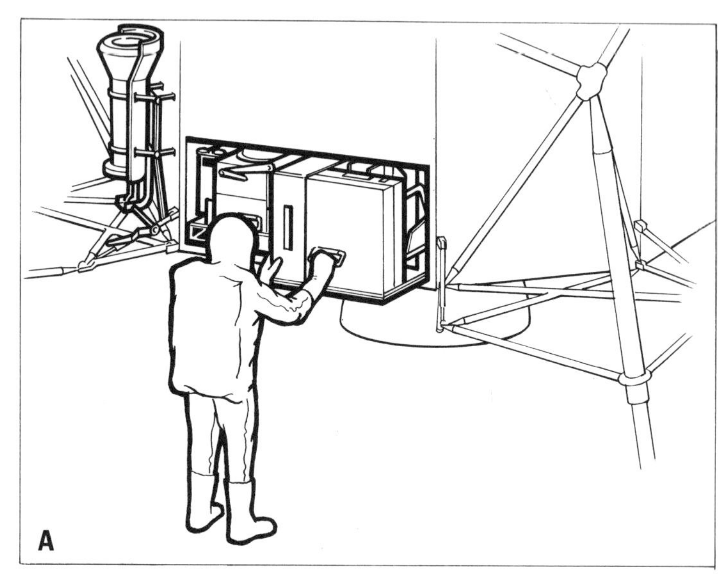

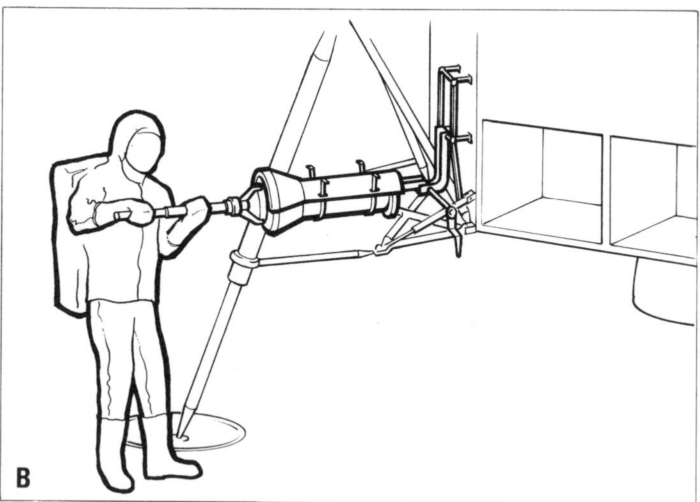

Deployment sequence of the **Apollo Lunar Surface Experiments Package (ALSEP)** described on pages 197–200. The SNAP-27 radioisotope generator is developed by the General Electric Company of the USA under the direction of the US Atomic Energy Commission. A. Apollo astronaut removes ALSEP components from stowage in base of lunar module. B. Radioisotope fuel capsule is removed from storage cask by hinging cask into

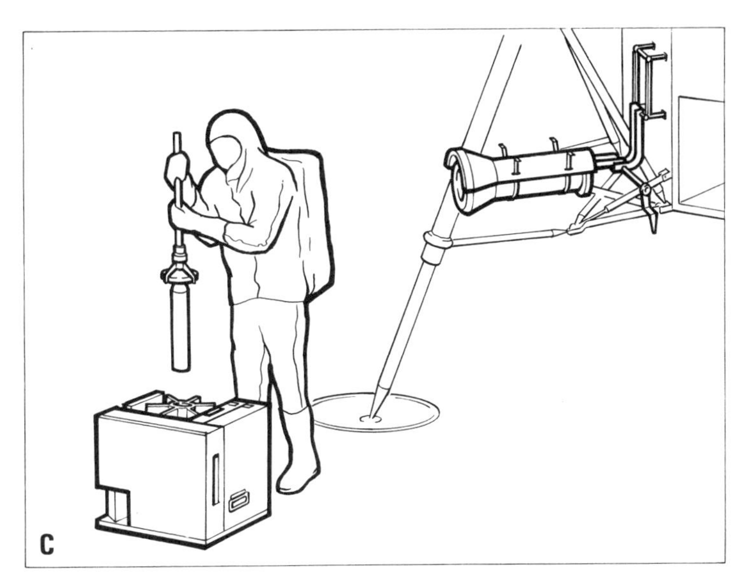

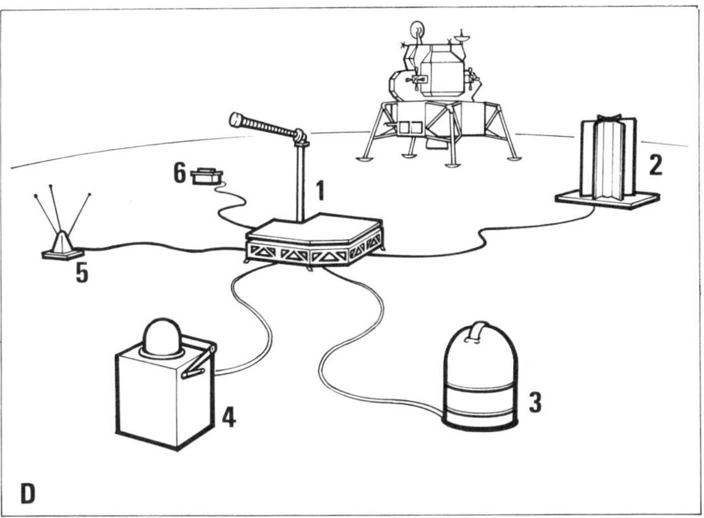

a horizontal position and engaging fuel capsule with handling tool. C. Fuel capsule is inserted into the SNAP-27 generator and locked into position. Full power operation occurs some 90 min. later. D. ALSEP components deployed to operate unattended in light and darkness for about one year. *Key:* 1. Aerial, data management and power distribution box; 2. SNAP-27 generator; 3. Seismometer; 4. Solar wind meter; 5. Magnetometer; 6. Ion detector.

upper and lower deck and end closure bulkheads, provided the main support structure. The space between the beam inter-sections formed the centre compartment which contained the throttleable descent engine. Maximum thrust was 10,500 lb.

Surrounding the engine were four propellant tanks, two oxidant tanks being located between the vertical beams and two fuel tanks between the horizontal beams. In the diagonal bays, adjacent to the propellant tanks, were helium, oxygen, and water tanks, batteries, lunar surface antennae, and scientific equipment.

The cantilever landing gear was mounted on outriggers that extended from the end of each of the two pairs of transverse beams. Each of the four legs consisted of a primary strut and footpad, a 'drive-out' mechanism, two secondary struts, two down-lock mechanisms, and a truss. All struts had crushable attenuator inserts as a means of absorbing impact loads at touchdown.

The landing legs remained folded until the spacecraft arrived in lunar orbit. The landing gear uplocks were then explosively released and springs in each drive-out mechanism extended the landing gear. Once extended each leg was locked in place by the two down-lock mechanisms.

At the time of lunar lift-off (or 'abort' before lunar landing) the two stages were separated by firing four explosive bolts at the stage interconnection. At the same time interstage wiring was explosively severed and other connections mechanically released.

APOLLO DIGEST (Lunar Programme)

Summary of Apollo flights including the first manned excursions to the Moon.

MISSION DESIGNATION AND DATE	DESCRIPTION
AS-201 26 Feb. 1966	Unmanned, sub-orbital space vehicle development flight by Saturn 1B; demonstrated space vehicle compatibility and structural integrity including command module's heat shield qualification for Earth orbital re-entry speeds.
AS-203 5 July 1966	Unmanned, orbital launch vehicle development flight by Saturn 1B; demonstrated second stage restart and cryogenic propellants storage in zero g conditions. Liquid hydrogen pressure test.
AS-202 25 Aug. 1966	Unmanned, sub-orbital space vehicle development flight by Saturn 1B; demonstrated structural integrity and compatibility; command module's heat shield performance.
Apollo 4 9 Nov. 1967	First Apollo/Saturn 5 flight, unmanned. Earth orbital to 18,075 km. (11,234 miles) apogee, space vehicle development flight. Demonstrated Saturn 5 rocket performance and command module's heat shield for lunar mission re-entry speeds.
Apollo 5 22–23 Jan. 1968	First Apollo lunar module flight by Saturn 1B, unmanned, Earth orbital. Demonstrated spacecraft systems performance, ascent and descent stage propulsion firings and restart, and staging.

Apollo 6 4 April 1968	Second flight of Saturn 5, unmanned, Earth orbital, launch vehicle development. Demonstrated Saturn 5 rocket performance and Apollo spacecraft sub-systems and heat shield performance.
Apollo 7 11–22 Oct. 1968	First manned Apollo flight with astronauts Schirra, Eisele, and Cunningham, by Saturn 1B; Earth orbital, demonstrated spacecraft, crew and support element performance. Highly successful flight lasted 10 days 20 hours, including eight service module main engine firings. First live TV from manned craft.
Apollo 8 21–27 Dec. 1968	History's first manned flight from Earth to another body in Solar System, by astronauts Borman, Lovell, and Anders, launched by Saturn 5; included 10 orbits of the Moon (20-hr. period) and safe return to Earth, TV and photography of Moon and Earth by astronauts.
Apollo 9 3–13 Mar. 1969 CSM 'Gumdrop' LM 'Spider'	First manned flight with complete Apollo spacecraft, including lunar module, by astronauts McDivitt, Scott, and Schweickart, into Earth-orbit by Saturn 5. First Apollo EVA, rendezvous and docking between LM and CSM, live TV, photo-observation of Earth, etc.
Apollo 10 18–26 May 1969 CSM 'Charlie Brown' LM 'Snoopy'	Second manned lunar orbit mission including descent of LM to within 15,240 m. (50,000 ft) of lunar surface, with astronauts Stafford, Young, and Cernan, launched by Saturn 5. Lunar module demonstrated full operational capabilities near the Moon short of actual landing.
Apollo 11 16–24 July 1969 CSM 'Columbia' LM 'Eagle'	First manned landing on Moon achieved by Neil Armstrong and Edwin Aldrin, with Michael Collins in orbiting Apollo parent. Confirmed touchdown point in Sea of Tranquillity 0°41′51″N latitude, 23°26′E longitude. Seismometer, laser reflector and 'solar wind' collector deployed and 20·8 kg. (46 lb) of soil and rock samples returned to Earth. Total stay time on lunar surface 21 hr. 36 min. Mission lasted 195 hr. 18 min. 35 sec. Launch vehicle Saturn 5.
Apollo 12 14–24 Nov. 1969 CSM 'Yankee Clipper' LM 'Intrepid'	Second manned lunar landing achieved by astronauts Charles Conrad, Jr, and Alan L. Bean, with Richard F. Gordon, Jr, in orbiting Apollo parent. Landed in Ocean of Storms 2°45′S latitude, 23°34′W longitude, within 200 m. (656 ft) of Surveyor 3 moonprobe which arrived on 19 April 1967. Set out first ALSEP station. Returned TV camera and other small parts from Surveyor plus 34·1 kg. (75 lb) of rock and soil samples. Total stay time on lunar surface 31 hr. 31 min. Mission lasted 244 hr. 36 min. 25 sec. Launch vehicle Saturn 5.
Apollo 13 11–17 April 1970 CSM 'Odyssey' LM 'Aquarius'	Third manned landing mission; abandoned short of objective Fra Mauro formation. Astronauts James A. Lovell, Jr, John L. Swigert, and Fred W. Haise. Oxygen tank of SM fuel-cells exploded when craft was some 205,000 miles from Earth approaching the Moon. Safe return achieved after circumnavigation of Moon, using lunar module oxygen and electrical supplies. Launch vehicle Saturn 5.
Apollo 14 31 Jan.–9 Feb. 1971 CSM 'Kitty Hawk' LM 'Antares'	Fourth manned landing mission achieved by astronauts Alan B. Shepard, Jr, and Edgar D. Mitchell, with Stuart A. Roosa in orbiting Apollo parent. Landed at Fra Mauro 3°40′S latitude, 17°28′W longitude, within 18·3 m. (60 ft) of intended target. Set out second ALSEP station. Used hand cart to carry cameras and geological tools. Shepard practised golf shot. Obtained 42·6 kg. (94 lb) of lunar samples including two rocks weighing about 4·5 kg. (10 lb) each. Total stay time 33 hr. 31 min. Mission lasted 216 hr. 1 min. 57 sec. Launch vehicle Saturn 5.
Apollo 15 26 July–7 Aug. 1971 CSM 'Endeavour' LM 'Falcon'	Fifth manned landing mission achieved by astronauts David R. Scott and James B. Irwin, with Alfred M. Worden in orbiting Apollo parent. Landed at Hadley Apennine 26°6′N latitude, 3°39′E longitude. First Apollo 'J'-class mission carrying Lunar Roving Vehicle. Set out third ALSEP station. Obtained 76·6 kg. (169 lb) of rock and soil samples. First live TV coverage of LM lift-off from Moon. Total stay time 66 hr. 55 min. Mission lasted 295 hr. 11 min. 53 sec. Released sub-satellite in lunar orbit. Launch vehicle Saturn 5.
Apollo 16 16–27 April 1972 CSM 'Casper' LM 'Orion'	Sixth manned landing mission achieved by astronauts John W. Young and Charles M. Duke, Jr, with Thomas K. Mattingly, II, in orbiting Apollo parent. Landed at Descartes highlands 8°54′S latitude, 15°30′E longitude, with second Lunar Roving Vehicle. Set out fourth ALSEP station. Three EVA's. Obtained 95·2 kg. (210 lb) of rock and soil samples. Released sub-satellite in lunar orbit. Total stay time 71 hr. 6 min. Mission lasted 265 hr. 51 min. 05 sec. Launch vehicle Saturn 5.
Apollo 17 7–17 Dec. 1972 CSM 'America' LM 'Challenger'	Seventh and last manned landing mission of series, achieved by Eugene A. Cernan and Harrison H. Schmitt, with Ronald E. Evans in orbiting Apollo parent. Landed at Taurus Littrow 20°10′N latitude, 30°45′E longitude, with third Lunar Roving Vehicle. Set out fifth ALSEP station. Three EVA's. Obtained 116·5 kg. (257 lb) of rock and soil samples. Released sub-satellite in lunar orbit. Total stay time 74 hr. 58½ min. Mission lasted 301 hr. 51 min. 59 sec. Launch vehicle Saturn 5.

Skylab

Skylab, America's first manned space-station, was launched by a two-stage Saturn 5 rocket on 14 May 1973 into an orbit which ranged between 427 and 439 km. (265 and 273 miles) inclined at 50° to the Equator. Between mid-1973 and February 1974 three Apollo CSM's launched by Saturn 1B vehicles allowed three-man crews to spend periods of about 28, 59, and 84 days on board.

Skylab and the Apollo CSM combined had an overall length of 53·7 m. (118·5 ft) and a volume of nearly 390 m.^{3} (13,000 ft^3). The total mass was approximately 86,275 kg. (190,200 lb).

Further details of the station will be found in the companion volume *Frontiers of Space*. Living conditions on board were more normal than anything that had been achieved in space before. As big as a medium-size house with about 50 times more volume than the Apollo CSM, there were proper washing, toilet, and sleeping facilities. Astronauts wore soft shoes, cotton undergarments, light jackets, and slacks. Lockers provided 60 changes of jackets, shirts, and slacks. There were 210 pairs of pants, 15 pairs of boots and gloves, 55 bars of soap, 95 kg. (210 lb) of towels, and 1,800 urine and fecal bags – even a vacuum cleaner.

'Luxuries' included music broadcasts from Earth, an entertainment kit with tape recorder and microphones, headsets, cassettes, playing cards, books, balls, and binoculars (for looking at the Earth and the heavens through the 45·7-cm. (18-in.) viewing window in the wardroom).

The research emphasis throughout was upon practical human benefits, the number of different experiments and engineering investigations totalling about 270 including 54 different items of experimental hardware. Some of the most important studies were related to astronomy, biology, medicine, processing of materials under weightless conditions, and Earth observation. Earth-resource observations included such 'down-to-Earth' topics as pollution of air and water, flooding, erosion, weather, crop deterioration, and the discovery of new sources of minerals and fossil fuels.

The Apollo Telescope Mount, with its cruciform solar panels, was used to observe and record the behaviour of the Sun using an array of UV and X-ray instruments. The results greatly increased knowledge of the Sun and the multitude of solar influences on the Earth's environment.

Earth observations using the Earth Resources Experiment Package (EREP) included a multispectral photographic facility, an IR spectrometer, a 10-band multispectral scanner, microwave scatterometer, and an L-band radiometer.

Looking towards the day when 'Space Factories' orbit the Earth, the Skylab astronauts experimented with industrial processes which may be enhanced by the unique weightless, vacuum environment of space. The M512 material processing apparatus included a spherical vacuum work chamber, a small electron gun (2 kW), and controls. Using the electron gun as a heat source, it was possible to observe the flow of molten metals under weightless conditions. (The Russians had previously tested electron beam welding during the flight of Soyuz 6.)

Sphere-forming experiments were also conducted in vacuum and zero-g. In the M518 multipurpose electric furnace, it was possible to test composite casting techniques – also the production of large crystals of high purity, e.g. germanium and indium-antimonide.

The crews subjected themselves to the most extensive medical experiments ever carried out in space, principally to determine the effects of long exposure to weightlessness and re-adaptation to normal gravity conditions. A regular physical training programme developed by the NASA doctors proved most beneficial. These carefully monitored experiments increased man's knowledge about the working of his own body – a major contribution to medical science.

In addition to the professional experiments, NASA threw open Skylab to student ideas. Science classes were especially quick to grasp the value of re-studying under weightless conditions above the atmosphere classical problems in chemistry, physics, and biology. Of more than 3,400 proposals received from US secondary schools, the space agency selected 25 prize-winners. Proposals included seeing if a spider could spin a web while weightless; seeing how a fertilized chicken egg develops; growing radishes to see if the roots still grow downwards towards the Earth's centre and the stem upwards, and photographing volcanoes using heat sensing infra-red film as a possible advance warning of eruptions. Nineteen of the prize-winners actually had their ideas incorporated in the Skylab programme.

Skylab at the Cape

For the Skylab project NASA adapted Launch Complex 39 which had served to launch the Apollo moonships (page 60). The station itself ascended on the nose of a two-stage Saturn 5 from LC 39A, the three Apollo CSM's on Saturn IB launchers from LC 39B.

As before the launch vehicles were assembled in the VAB on the transportable steel structure called the Mobile Launcher. Following check-out the vehicles were moved to the launch area on their mobile launch platforms by the crawler-transporter.

Although few modifications to standard equipment were required to launch Skylab itself, major changes were necessary to accommodate the smaller Saturn IB/Apollo CSM which carried the astronauts. They included providing a 39 m. (128 ft) tall structural steel pedestal to raise the launch vehicle to a position compatible with the appropriate service arms of the Mobile Launcher.

The same equipment was subsequently used to launch the Apollo CSM of the Apollo-Soyuz Test Project (ASTP).

Skylab Apollo CSM

To launch three astronaut teams to the Skylab space-station NASA adapted command and service modules which had been built for the Apollo Moon programme. As these craft were not required to travel as far as their Moon-bound sisters, two service propulsion system tanks were eliminated. On the other hand, it was necessary to add more re-entry control system propellant to provide a back-up de-orbit capability.

Several changes were also made to the command module control panels, and the design was modified so that two extra couches could be put into the command module, should it be necessary to fly a rescue mission. (In this event, the craft would have a two-man flight crew, allowing full accommodation for three rescued astronauts.)

Additional stowage capacity was provided by increasing locker size required to carry various supplies to Skylab, and for returning 'hard data' such as films and tapes to Earth. Two experiments – SO71 and SO72 – were also housed in the command module. These were aimed at determining the physical functions of pocket mice and vinegar gnats when removed from gravity and a 24-hour geophysical environment.

Finally, as one side of the Apollo CSM was in the shade during docked manœuvres with Skylab, with temperatures approaching $-100°$ F., the craft's thermal control system was improved by the addition of thermostatically controlled heaters.

Space Shuttle (pages 94–6)

The NASA Space Shuttle, to be introduced in the early 1980s, is designed drastically to reduce the cost of space operations by recovering the major part of the launch vehicle for re-use. Lifting off like a rocket, it operates in orbit as a spacecraft and returns to land horizontally on a runway like an aircraft. Shuttles will carry up to 29,474 kg. (65,000 lb) of cargo, four crew-members and up to four passengers into a close Earth-orbit. They will *return* up to 14,515 kg. (32,000 lb) of cargo to Earth.

The Shuttle system basically has three main elements: the payload-carrying orbiter, an external tank containing liquid hydrogen, liquid oxygen propellants for the orbiter's main engines, and two solid rocket boosters, which separate at 43 km. (27 miles) altitude when the velocity has reached 5,170 km./hr. (3,210 m.p.h.), and are designed for parachute recovery in the Atlantic. It is hoped to recover the boosters for refurbishment and re-use in later missions. The external tank jettisons just before the point of orbital insertion and impacts in the ocean.

Details of the propulsion systems used in the Shuttle are given in the companion volume *Missiles and Rockets*.

The orbiter itself (page 96) is about the size of a medium-range jetliner. It has a blended wing-fuselage design for optimum aerodynamic and manœuvring characteristics; and despite the wide range of its performance – including re-entry from orbit – the structure employs aluminium alloy as the basic material. Thermal protection consists of a silica fibre-based high-temperature and low-temperature re-usable surface insulation over most of the vehicle, and a reinforced carbon-carbon composite for the nose and wing leading edges.

The crew compartment has seating for two pilots and two mission crewmen on the top deck, with duplicate flight controls for pilot and co-pilot. In the lower deck is space for up to six passengers (male and female) and basic amenities, including kitchen and toilet facilities. Aft of the crew compartment is the 18·3 m. (60 ft) by 4·6 m. (15 ft) diameter cargo bay. In this can be fitted a manipulator arm equipped with television for deploying and retrieving one or more payloads. Alternatively, the cargo bay can accommodate a fully equipped European Space Laboratory (see the companion volume *Frontiers of Space*, 3rd edition).

The Environmental Control and Life Support Systems (EC/LSS) comprise: (1) the Atmospheric Revitalization Sub-system (ARS); (2) the Food, Water, and Waste Sub-system (FWWS); (3) the Active Thermal Control Sub-system; and (4) the Airlock Support Sub-system (ASS).

Avionics includes the following systems: (a) guidance, navigation, and control (GN&C); (b) data processing and software; (c) communications; (d) displays and controls; (e) flight instrumentation; and (f) fuel-cells electrical power distribution.

Shuttle flight profile: Orbit insertion and circularization: altitude (typical) 185 km. (115 miles); velocity 28,300 km./hr. (17,600 m.p.h.). Orbit operations: altitude 161–966 km. (100–600 miles); duration 7–30 days. Atmospheric entry: altitude 122 km. (76 miles); velocity 28,100 km./hr. (17,466 m.p.h.). Manœuvrability capability: crossrange ±1,460 km. (±905 miles) from entry path; landing velocity 346 km./hr. (215 m.p.h.).

Preparing for the Space Shuttle

The day of the big Saturn boosters at the Cape finally drew to a close with Skylab and the Apollo-Soyuz Test Project in 1973–5 – with preparations already under way to accommodate the next major advance, the re-usable Space Shuttle.

Despite the revolutionary nature of this project, NASA kept modifications to major facilities at the Cape to a minimum. The existing Mobile Launchers were again adapted to enable the Shuttle to be assembled complete in the VAB before it was moved by the crawler-transporter to Launch Complex 39.

New facilities included a 4,572 m. (15,000 ft) runway, with 305 m. (1,000 ft) of overrun at each end, to recover the winged orbiter after its mission and a Maintenance and Check-out Facility near the end of the runway, linked by roads and aprons, making it possible to service and refurbish the vehicle, and generally return it to flight status.

With a new payload fitted it would then be towed back to the VAB to be vertically mounted on the Mobile Launcher and fitted with a new drop tank and solid rocket boosters (see the companion volume *Missiles and Rockets*).

Before operational testing began, NASA arranged to fly the Shuttle orbiter prototype on the back of a specially modified Boeing 747. The arrangement was reminiscent of the 1938 Short-Mayo Composite aircraft in which a large flying boat, *Maia*, was used to launch in mid-air a smaller *Mercury* seaplane with tanks full for a trans-Atlantic flight.

The 747 scheme meant separating the orbiter – which had been fitted with a special tail fairing – from the mother aircraft with the orbiter set at an 8° angle of attack on structural attachments.

Flight tests of the prototype Orbiter 101 were scheduled to begin in 1977 at the Flight Research Center, Edwards Air Force Base, California, first with the orbiter captive on the 747 and then in free flight following release from the carrier aircraft at about 9,144 m. (30,000 ft) by the lift derived from its wings. This gave test pilots the opportunity to perform approach and landing manœuvres touching down on the Edwards Dry Lake runway.

Launch tests of the second prototype (102) from the Kennedy Space Center were expected to return the orbiter to Edwards because of the greater latitude available to pilots for returning the spaceplane, without main engine power, to a runway landing. The craft would then be air-lifted by the 747 back to Cape Canaveral.

GLOSSARY OF SPACE TERMS

Ablation	erosion of a solid body (e.g. a spacecraft's heat shield) by a high-velocity, high-temperature gas stream
'Abort'	emergency termination of a rocket launching
Acceleration	rate of change of velocity
Apogee	that point in a terrestrial orbit farthest from the Earth
Astronautics	the science and technology of spaceflight
Ballistic trajectory	path followed by vehicle in unpowered flight
Booster	propulsion unit used in initial stage of flight
Chaff	metal-foil strips ejected from a tracked space-vehicle to enhance the radar response
Cosmonaut	Soviet term for an astronaut; a space-traveller
Cryogenic propellant	one that, at atmospheric pressure, has a boiling-point below 0° C of the liquefied gas (e.g. liquid oxygen, liquid hydrogen)
Escape velocity	velocity needed to escape from a given point in a gravitational field (e.g. 6·95 miles/sec. to escape from the Earth's surface)
Free-fall	motion of a body moving in a gravitational field without power (see *weightlessness*)
Fuel-cell	battery in which chemical reaction is used directly to produce electricity
g	symbol for the acceleration of a freely moving body due to gravity at the surface of the Earth
Gimbal	mechanical frame for a gyroscope or power-plant, usually with two perpendicular axes of rotation

Hypergolic	term applied to a fuel and oxidant which ignite spontaneously on contact
Lunar	of or pertaining to the Moon
Mach number	ratio of the speed of a vehicle to the local speed of sound, approximately 750 m.p.h. at sea-level
Multi-stage rocket	a rocket having two or more stages which jettison sequentially
Orbit	path of a body relative to its primary
Orbital period	time taken by an orbiting body to complete one orbit of its primary
Orbital velocity	speed of a body following closed or open orbit, generally applied to elliptical or near circular orbits, e.g. close to Earth about 18,000 m.p.h.
Perigee	that point in a terrestrial orbit closest to the Earth
Probe	unmanned vehicle sent into space to gather information by means of instruments
Propellant	liquid or solid substance or substances burned in a rocket engine to produce thrust
Re-entry	return of a space vehicle into the Earth's atmosphere
Rendezvous	planned meeting between two spacecraft in orbit
Retro-rocket	rocket fired against direction of motion to reduce speed
Shingles	thin interleaved plates of metal used as outer heat-resistant skin
Solar	of or pertaining to the Sun
Solar cell	silicon wafer that converts sunlight directly into electricity
Solar wind	constant stream of ionized gas atoms, mostly hydrogen, from the Sun
Stage	self-contained propulsion system in a launch vehicle which has stages operating sequentially

Telemetry system for relaying data from a spacecraft's instruments by radio to a ground receiving station

Terrestrial of or pertaining to the Earth

Umbilical flexible connector which conveys power to a spacecraft and/or launch vehicle before take-off; connector supplying an astronaut during extra-vehicular activity

Ullage that volume in a tank not occupied by propellant

Ullage rocket propulsion unit applied under weightless conditions to force propellant to outlets in the bottom end of a tank, for purposes of engine re-starting, by reactive effect

Weightlessness state experienced in ballistic flight – in orbit or free-fall – when a body experiences no mechanical stress because the gravitational attraction is opposed by equal and opposite inertial forces

Zero-g condition of weightlessness

INDEX

Numerals in **bold** refer to illustrations